International Perspectives on Publishing Platforms

With large-scale scholarly projects dedicated to digitizing print-based magazines and a concurrent turn towards digital mapping and data visualization, periodicals that were once accessible only in the archive now have the capacity to reach a wider audience, and make visible previously overlooked networks and connections enacted within and across the magazines. *International Perspectives on Publishing Platforms: Image, Object, Text* offers a unique contribution to the field of periodical studies, while also broadening the scope of purview to consider related content with regards to other relevant printed matter and cultural products, as well as digital archiving strategies.

Including interdisciplinary contributions from academics around the world, the volume presents a wide range of approaches to periodicals and printed matter from Africa, Asia, Europe, the Middle East, Latin America, and the Caribbean. Questions of material print culture and the digital realm are considered both via theoretical and more empirical approaches. As a whole, the book considers the pluralism of perspectives that the study of periodicals and printed matter contribute to our historical understanding of various political and social issues, and also devotes attention to the ways in which digital archiving projects can be instrumentalized as a strategy for filling in gaps in the historical record.

International Perspectives on Publishing Platforms should be of great interest to researchers, academics, and postgraduates engaged in the study of periodicals, publishing, book history, world literature, digital humanities, media, visual and material culture.

Meghan Forbes is currently the C-MAP (Contemporary and Modern Art Perspectives) Fellow for Central and Eastern Europe at The Museum of Modern Art in New York and a Visiting Scholar at the Institute for Public Knowledge at New York University. Previously, she was Czech Lecturer in the Department of Slavic and Eurasian Studies at the University of Texas at Austin. She holds a PhD from the Department of Slavic Languages and Literatures at the University of Michigan, Ann Arbor where she completed her dissertation, "In the Middle of It All: Prague, Brno, and the Avant-Garde Networks of Interwar Europe." She is the recipient of numerous grants for her research, including an IIE Fulbright award to Berlin in the 2014–2015 academic year.

Digital Research in the Arts and Humanities
Series Editors: Marilyn Deegan, Lorna Hughes, Andrew Prescott and Harold Short

Digital technologies are increasingly important to arts and humanities research, expanding the horizons of research methods in all aspects of data capture, investigation, analysis, modeling, presentation, and dissemination. This important series covers a wide range of disciplines with each volume focusing on a particular area, identifying the ways in which technology impacts specific subjects. The aim is to provide an authoritative reflection of the "state of the art" in technology-enhanced research methods. The series is critical reading for those already engaged in the digital humanities, and of wider interest to all arts and humanities scholars.

Critical Gaming: Interactive History and Virtual Heritage
Erik Champion

Literary Mapping in the Digital Age
Edited by David Cooper, Christopher Donaldson and Patricia Murrieta-Flores

Historic Newspapers in the Digital Age: "Search All About It!"
Paul Gooding

Cultural Heritage Communities: Technologies and Challenges
Edited by Luigina Ciolfi, Areti Damala, Eva Hornecker, Monika Lechner and Laura Maye

Generative Systems Art: The work of Ernest Edmonds
Francesca Franco

Cultural Heritage Infrastructures in Digital Humanities
Edited by Agiatis Benardou, Erik Champion, Costis Dallas and Lorna Hughes

The Strehlow Archive: Explorations in Old and New Media
Hart Cohen

Shakespeare's Language in Digital Media
Edited by Janelle Jenstad, Mark Kaethler and Jennifer Roberts-Smith

New Directions in Mobile Media and Performance
Camille Baker

The Shape of Data in the Digital Humanities: Modeling Texts and Text-based Resources
Julia Flanders and Fotis Jannidis

For more information about this series, please visit: www.routledge.com/history/series/DRAH

International Perspectives on Publishing Platforms

Image, Object, Text

Edited by Meghan Forbes

Routledge
Taylor & Francis Group
LONDON AND NEW YORK

First published 2019
by Routledge

2 Park Square, Milton Park, Abingdon, Oxfordshire OX14 4RN
52 Vanderbilt Avenue, New York, NY 10017

Routledge is an imprint of the Taylor & Francis Group, an informa business

First issued in paperback 2020

British Library Cataloguing-in-Publication Data
A catalogue record for this book is available from the British Library

Library of Congress Cataloging-in-Publication Data
A catalog record has been requested for this book

ISBN: 978-1-138-58966-7 (hbk)
ISBN: 978-0-367-66283-7 (pbk)

Typeset in Times New Roman
by Apex CoVantage, LLC

Contents

Figures

Contributors

Nana Ariel is a writer, researcher, and lecturer at Tel Aviv University, Israel, and was formerly a Fulbright post-doctoral Visiting Scholar at Harvard University. Her studies and essays, published in English, French, and Hebrew, center around rhetoric, literature, language, and material culture. Her recent studies explore the rhetoric of cultural movements in the international and Hebrew context, the role of publishing devices, such as manifestos and little magazines, and specifically the tensions between originality and repetitiveness. She is also engaged with the exploration of interactions between theoretical and practical knowledge in humanistic pedagogy and rhetoric.

Jennifer Bajorek writes and does research on literature, philosophical aesthetics, and photography. Her articles on photography, photographic archives, and contemporary institutions for photography in Africa have appeared in *Aperture Magazine*; *Autograph*; *Theory, Culture & Society*; *Third Text*; *Social Text*; *Africultures*; *Afriphoto*; *Fotota*; and the Galerie du Jeu de Paume blog. She is an assistant professor of Comparative Literature at Hampshire College, USA, and a research associate in the Research Centre in Visual Identities in Art and Design in the Faculty of Art, Design, and Architecture at the University of Johannesburg, South Africa. She will be a residential research fellow at the Clark Art Institute in 2019. Her latest book, *Unfixed: Photography and Decolonial Imagination in West Africa*, is forthcoming.

Olympia Bhatt completed her PhD from the School of Arts and Aesthetics, Jawaharlal Nehru University, New Delhi, India. Her research interests include history of sound technology in India, visual and aural ethnography, and media archaeology, as well as new ways of contemporizing and engaging the archive in this pervasive digital present.

María A. Cabrera Arús, PhD (Sociology, New School for Social Research, USA), studies the impact of fashion and domestic material culture on regime stability and legitimation, especially in the Caribbean, during the Cold War years. Her research has been published in the peer-reviewed journals *Kamchatka, Theory & Society* and *Cuban Studies*, as well as in book anthologies. Cabrera Arús is the author of the New Challenge Award for Social Innovation–winning project

Cuba Material, a virtual archive of Cuban material culture from the Cold War era, and co-curated the exhibitions *Pioneros: Building Cuba's Socialist Childhood* in the Sheila C. Johnson Design Center at Parsons School of Design (September 17–October 1, 2015) and *Cuban Finotype and Its Materiality*, at *Cabinet* magazine (October 21, 2015). She was the 2016–2017 Postdoctoral Research Fellow in the Sawyer Seminar "Cuban Futures Beyond the Market" at the King Juan Carlos I of Spain Center of New York, and currently teaches at Gallatin School of Individualized Study, New York University.

Zanna Gilbert is a research specialist in the curatorial department of the Getty Research Institute, USA. She completed her PhD at the School of Philosophy and Art History at the University of Essex, UK, in collaboration with Tate Research. From 2012 to 2015, Gilbert was the Andrew W. Mellon Postdoctoral Fellow in the Department of Drawings and Prints at The Museum of Modern Art (MoMA) in New York City, where she was responsible for C-MAP research focusing on modern and contemporary art in Latin America and was co-editor of the online publication "post." She has curated a number of exhibitions, including *The Unmaker of Objects: Edgardo Antonio Vigo's Marginal Media* (MoMA, 2014) and *Home Archives: Paulo Bruscky and Robert Rehfeldt's Mail Exchanges* (Chert, Berlin, 2015); she was co-curator of the Getty's PST: LA/LA exhibition *Making Art Concrete: Works from Argentina and Brazil in the Colección Patricia Phelps de Cisneros*.

Naomi Kuromiya is a PhD student at Columbia University, USA, where she studies modern Japanese art history in the Department of Art History and Archaeology. Her research interests include the postwar intersections of "avant-garde" and "tradition" in Japan and artistic interactions between Japan, USA, and France. She completed her MA at the Institute of Fine Arts, New York University, with a thesis that examined the artistic influence of the postwar Japanese calligraphy collective, Bokujin-kai. She has also worked for the Lyonel Feininger Project LLC and for The Museum of Modern Art in New York, where she held the 2012–2014 Dedalus Fellowship in the Museum Archives.

Hannah Pröbsting is an independent researcher and museum professional, working at the Musée de l'Elysée Museum of Photography in Switzerland. Her PhD dissertation "Winning New Freedom: Intersections of Text and Image in the Arts of Kurt Schwitters" (University of Michigan, Ann Arbor, 2017) brought together the theoretical essays of German avant-garde writer, artist, and designer Kurt Schwitters (1887–1948) with his fine and applied art to examine three sites where text and image converge – typography, graphic design, and collage arts.

Camilla Salvaneschi is a doctoral candidate in Visual Culture at the Università Iuav di Venezia, Italy, and at the University of Aberdeen, UK. Her thesis analyzes the art magazines published by recurring exhibitions such as Documenta, Manifesta and the Venice Biennale. Her research interests include the evolution of contemporary art magazines and the role of magazines and criticism in the art system. She co-curated with Mario Lupano and Saul Marcadent the exhibition and related conference *Fiamme. Fifty Contemporary Art Magazines: Art, Fashion, Architecture, and Design* (Università Iuav di Venezia, 2017).

Kenan Tekin is an Assistant Professor at Yalova University, Turkey. He holds a PhD from the department of Middle Eastern, South Asian, and African Studies at Columbia University, with a dissertation entitled "Reforming Categories of Science and Religion in the Late Ottoman Empire." Previously, he earned an MA in Religion from Duke University, USA. Tekin's research interests include *longue durée* study of Ottoman intellectual thought, history, and philosophy of science and religion. Tekin is also interested in the transition from manuscript to print culture, and the social and cultural impacts of that transition on modernization and secularization processes, as well as the genealogies of modern discourses.

Introduction[1]

Meghan Forbes

In the first semester after completing my PhD, I had the opportunity to teach my dream course: a class on avant-garde Central and Eastern European magazines of the period between the two world wars. In preparing the syllabus, I aimed to incorporate readings that could at least tangentially connect to wider networks of avant-garde magazine production outside of my own geographical area of expertise. While it was no trouble at all to find sources on Western European periodicals – such as histories related to Dada, Surrealism, and Italian Futurism – as well as on the culture of the Anglo-American "little magazine," it was far more difficult to find secondary sources relating to artistic and literary developments happening outside of a Western, white canon. I was further frustrated in my efforts to bring these perspectives into the course via visits to the campus's renowned research library, through which I hoped to draw on a more diverse sampling of interwar primary source materials; not only was there a notable gap in the library's holdings of the period that did not pertain to those movements already mentioned above, so, too, did it almost entirely lack materials directly associated with the Central and Eastern European focus of the course – for instance, magazines from Czechoslovakia, Germany, Poland, Hungary, and Yugoslavia – that were in clear conversation with items held in the collection.

Working within the US, my difficulty in tracking down the sources that I wanted to teach points toward a multipronged problem that indeed many scholars, librarians, and archivists have dedicated themselves to making more visible, and thereby working to correct – often from within the very institutions that tend to perpetuate collection and research practices that make wide these gaps. It is also the impetus for the present volume. A relative paucity of English-language scholarship on print and publishing histories that are outside of an inscribed Anglo-American modernist canon is arguably both a result of, and a reason for, the dearth of primary source documents held in libraries and archives. And certainly it is an issue that cannot be solved by libraries and archives – and librarians and archivists – alone, though an active engagement by these communities is of course essential.[2] It must also be addressed through the research, writing, and teaching of scholars, and greater opportunity for the direct engagement of diverse publics with institutions that are meant to serve them. New initiatives in digital archiving and open-access online collections provide expansive new possibilities in this regard that allow archivists, librarians, researchers, students, and the general public greater opportunity for direct engagement and collaboration, several examples of which are highlighted in this volume.

International Perspectives on Publishing Platforms: Image, Object, Text points to the breadth of materials beyond those that have tended to occupy the most space in collections and enjoy the most prominence in the Western field of periodical studies, that is, early twentieth-century English-language magazines on an American–UK axis.[3] The nine chapters that follow are meant together to offer a broadly international range of materials that represent exciting contributions to specific histories of print culture and reflections on contemporary archival practices. These span in scope from a discussion of the role of the printing press in the late nineteenth-century Ottoman Empire, post–World War II mail art networks between Latin America and Europe, magazine reproductions of Japanese calligraphy that resulted in international exhibitions, to present-day digital archiving strategies of periodicals in Asia and Africa. In short, this book offers the very plurality of texts I would hope to teach.

Through reading these chapters myself, and learning more about the topics they present, I have also been made aware of many other excellent sources and scholars. The bibliography to this book alone offers the reader an excellent resource in gaining a more transnational, comparative perspective on print cultures and archival strategies of the last 150 years. Growing attention towards periodicals and related print materials previously overlooked (by the dominating discourse) is palpable and is coupled with the emergence of digital publishing and archiving platforms that allow for marginalized items to become more visible to wider publics; it is at this overdue moment of cultural shift and exciting new scholarship that this volume enters.

However, the goal of *International Perspectives on Publishing Platforms* is not one of mere accretion, simply an expanding of the canon, so to speak. Rather it offers, when read as a whole, a theoretically and methodologically diverse approach to the periodical and related print culture, and engages associated questions with regard to collecting, preserving, and archiving. The periodical, especially when representing a history that is little known outside of its local context, can serve as an invaluable archive of that history as it was enacted in real time. It becomes a privileged site therefore of writing histories of non-Western modernities and modalities, and, in conjunction with the digital turn, the periodical-as-archive can be archived online, further making visible what it contains.[4] Thus it follows that a book that takes up the periodical and printed matter would also bear a consideration of the archive – referring both to the magazine object itself, and other objects besides, and that place at which these objects are stored.[5] And a volume ostensibly dedicated to "publishing platforms" in 2019 cannot fail to consider the interconnectivity of the physical and digital realms as means of publication, modes of distribution, and sites of storage.

Whereas previous scholarly manuscripts and edited volumes related to the field of periodical studies have tended to delineate an intentionally narrow temporal or thematic focus, or definition of what constitutes a periodical – such as the

English-language interwar "little magazines"[6] – the present volume dares to both expand the temporal range of objects under consideration and widen the definition of the form itself. A contemporary art practice that grows out of working at a family commercial printing press in Nigeria, the public display of unlikely advertising posters in Germany in 1927, and an open-access digital archive of material artifacts from Cold War–era Cuba all find a home in this volume, which as a whole presents a strong case for working across disciplinary, medial, and geographic borders.

What is to follow in this introduction is an outline of the historical and theoretical framework within which the present volume was conceived, alongside a simultaneous interrogation of the assumptions on which this relatively familiar narrative has been constructed in an attempt to burst the frame wide open, allowing for a less narrow, and ultimately more generative, discourse.

Technologies of and beyond print

Rapid developments in printing technologies in the late nineteenth century and their impact on the distribution of printed matter and the material output of various artists and intellectuals in the West are well documented. In *The Visible Word: Experimental Typography and Modern Art, 1909–1923*, Johanna Drucker writes, "The early twentieth-century avant-garde found no medium more suitable for both production and promotion of their formal innovations and activist agenda than the printed page. [. . .] Print media were affordable, available, and effective means of communication and public visibility."[7] The social impact of printed text in circulation at the turn of the twentieth century was interrelated with, and in fact dependent upon, concurrent major advancements in industrialization. Robert Scholes and Clifford Wulfman, in their *Modernism in the Magazines*, suggest that a

> major force in the shaping of modernity was technological, led by discoveries and improvements in transportation and communication, including the railroads and trolleys that assisted urbanization, along with new methods of printing words. [. . .] The mass media that first appeared in modernity were newspapers and magazines.[8]

Thus print culture became cheap and ubiquitous, more fully entering the everyday lives of ordinary people. And in the early decades of the twentieth century, ever new technologies were imagined and actualized that could offer an even wider, and more impactful, mode of information dissemination, such as film and television, the gramophone and radio.

In 1921, Karel Teige, a leading figure in the Czech interwar avant-garde, compared the invention of the printing press to that of the moving image in the essay "The Art of Today and Tomorrow," articulating the role the machine plays in reproducing, propagating, and even archiving art for the masses:

> It has been rightly said that the discovery of the cinema is for us as important as the discovery of the printing press for the people of the Renaissance: here,

> too, machine production disseminates art to its viewers. And what's more: the heroes of cinema projected photographically on the screen are captured for eternity, are immortal. And the vinyl record preserves the voice of its singers. – *Yes, all of modern artistic culture consists and must consist in mechanical production.*[9]

Teige's comparison of the development of cinema to that of the printing press, as the newest form of technological reproduction that brings information to a wide audience, and can likewise preserve that information beyond the lifespan of its active participants, reflects a broader interest in film as a new system of writing better suited to conveying information in fast-paced modernity. But simultaneously and conversely, it also hints at the ways in which early twentieth-century print practitioners aimed to productively engage newly available image and sound-based technologies to envision the ways in which print itself could be more dynamic, as for instance in the typographic work of Teige's Hungarian contemporary and instructor at the Bauhaus school in Germany, László Moholy-Nagy, exemplified in his "Dynamic of a Metropolis."

Much modernist print culture that pushed forward innovations in both textual and visual material was clearly influenced by, but not always generated through, the newest technologies of the print industry. Artists, authors, and small-scale publishers eager to take agency over their output often set up shop and brought the means of production into their own hands. In England, Virginia Woolf, with her husband Leonard, began operating Hogarth Press in 1917 with a hand press named after the house out of which they worked, and published some of the most important exemplars of English-language modernism, including an edition of T.S. Eliot's *The Waste Land* and Woolf's *A Room of One's Own*. And earlier, in the late nineteenth century, as Kenan Tekin shows in his chapter here on the Young Ottoman Ali Suavi's periodical publication the *Ulûm Gazetesi*, even limited and inconsistent access to print technologies – in this case in exile in Paris – could provide the conditions necessary to generate an independent, subversive political statement that could reach a substantial audience.

The implications of new forms and greater access to print technologies and modes of distribution can be observed around the globe. Drucker has suggested that "modern" print production is "only conceivable within the conditions of a highly industrialized culture, one in which mass print culture was the common currency of exchange of ideas and forms."[10] But at the same time, as this volume makes clear, there are numerous approaches to print and other media that operate entirely outside of a "highly industrialized" capitalist apparatus. For instance, Olympia Bhatt, in her discussion here on the first Indian talkies as represented in film magazines, brings to light ingenious approaches for making accessible the most modern technologies of the cinema in landscapes without the infrastructure typically assumed necessary for projection (such as electricity), with these machines then advertised in print.

Nevertheless, Western, Anglo-centric conceptions of modernism have far too often excluded cultural developments outside this circumscribed sphere in the writing of histories, the process of canon building, and exhibition curation.[11] Robin Kinross attempts to address this bias in the early pages of his *Modern Typography: An Essay in Critical History* by explaining that the "limitation of this book to the western world, and to typography employing Latin script" can be justified by the fact that "the overlap between 'modern' and 'western' is so great as to make them synonymous."[12] This is a wholly uncritical view of the modernist field, and one that disregards modernisms occurring elsewhere, in different places and at different times (and no doubt points to the limitations of the use of the term "modern" in and of itself, but this is a critique that cannot be engaged here).[13] It is also a modeling of, as Chana Kronfled describes it, "Euro-American geopolitics and linguistics [that] effectively leaves all that is not English, French, or German [. . .] outside our purview."[14] And Hannah Pröbsting's chapter for this volume shows how even in Germany – a major site of the sort of typographic production that interests Kinross – there were typographers who were invested in seeking solutions for an international script that pushed beyond received Latin orthographies. By bringing together in one place a wider range of examples of print culture, that represent an expansive language of type and approaches to print practices, it is the project of this volume to modestly correct this offhanded dismissal of non-Western print production as outside the purview of the "modern" world, as well as to trace many multilinguistic and transnational instances of network formation.

Kinross describes four "leading strands" in approaches to print history, the oldest being bibliographical history, followed by histories of printing technologies, and the more recent addition of cultural history that looks to the printing and publishing trades, and finally, the history of typography.[15] This volume tells a bit of all of these histories, but is in particular concerned with the capacity of print narratives to map various aspects of social and political life across a broad territory. It also pushes our notion of histories of "print" to consider the legacy of textual-visual production within the context of the digital turn. The then radical accessibility to printed matter and innovations in information dissemination at the turn of the twentieth century is unmatched by the proliferation of access to textual and visual data that has been enabled by the invention of the Internet, and its subsequent global impact. The field of periodical studies has also developed in tandem with these developments that seem to leave nearly no aspect of life untouched. It is therefore relevant and indeed imperative to ask what print histories might look like "beyond print."

This is a question that can be directed both towards born-digital publishing platforms and to historical print materials that are now available online. With large-scale scholarly projects dedicated to digitizing print-based magazines, and a concurrent turn towards digital mapping and data visualization, periodicals that were once sequestered in the archive now have the capacity to reach a wider audience, and make visible previously overlooked networks and connections enacted within and across the magazines. The Blue Mountain Project at Princeton University is a digital collection of magazines begun in 2012 that now houses 3,744 issues of "important periodicals of the European avant-garde."[16] Similarly, the Modernist Journals Project is a collaboration of Brown University and the

University of Tulsa that has been digitizing periodical collections since 1996, and includes in its holdings journals such as *Poetry*, *The Little Review*, *The Crisis*, and *The Egoist*.[17] At the University of Iowa, the International Dada Archive makes available digitally periodicals and pamphlets selected from a physical catalogue of approximately 60,000 related items.[18] All of these projects mark major advancements in open-access accessibility to collections, now available to researchers and an interested public with unrestricted Internet access anywhere around the world.

And while the above examples are mostly confined to iterations of early twentieth-century modernisms in the US, UK, and Western Europe (though the Blue Mountain Project does also include Czech, Polish, and Russian items), there are innumerable projects of small and large scale that account for other regions and time periods, such as the wiki Monoskop, initiated in Slovakia and an excellent resource for (in particular but by no means exclusively) the Central and Eastern European avant-gardes.[19] The Asia Art Archive, based in Hong Kong, aims to document "multiple recent histories of art in Asia,"[20] and the pan-African platform Chimurenga is dedicated to practices of self-archiving with its online library that "focuses on how we forge communities, produce and circulate knowledge and operate in the border zones between informal/formal, licit/illicit, chaotic/ordered, etc."[21] As many of the texts here bring to light, what was initiated at the local level can obtain international visibility and legibility through such digital platforms. And recent examples of publications that incorporate the digital from inception indicate that the question of where a periodical is sited and how it is catalogued becomes an integral part of the very practice of its publication.[22]

The essential archive

The present volume gives much attention to that archive within the archive – the magazine – but not exclusively so; it also makes space to encompass an engagement with the archive itself, and the essential role it plays in the facilitation (or obfuscation) of a cultural history that printed matter helps construct. A consideration of the physical and digital archive is foregrounded here, thus expanding the limits of the field of purview in considering the periodical and related material culture to also address how, where, and by whom these items are stored – an incredibly important, and yet somehow so often overlooked, site of knowledge production.[23]

The archive – from the large and federally controlled holdings of a given government to the carefully kept catalogue of items belonging to a single person and maintained by a private estate – can be conceived of as both a site of order and chaotic randomness, of visibility and invisibility, of inclusivity and exclusivity. If we imagine the physical archive as the Enlightenment's answer to the Wunderkammer, then we might see in it also, despite all semblance of order and cogent organization, some of its predecessor's chaos, "everything [. . .] so intertwined in a visible and invisible network."[24] The activation of the archive is possible through efforts to make publically accessible what is contained therein (whether that refers to a physical or digital space) and the compulsion of an individual to seek out this material and do something with it (whether in the form of scholarly writing, activist work, or art production, for instance).

Aleida Assmann points to the fact that getting objects into the archive is only half the battle, when she poses in her discussion of "Canon and Archive" the former term as active and the latter as passive, in relation to the work of memory. Assmann distinguishes between the canon as "actively circulated memory that keeps the past present" (as in a museum's collection of artworks that are on display) and the archive as "passively stored memory that preserves the past past" (as in that same museum's collection siloed in storage).[25] While the archive readily suggests a site in which objects are potentially forgotten with disuse, it can also harbor the potentiality that those items are reactivated through engagement.[26] Assmann situates the archive "on the border between forgetting and remembering; its materials are stored in a state of latency, in a space of intermediary storage."[27] In a sense, the archive exists as a placeholder for memory, so that we ourselves do not have to do the work of remembering everything; the objects in the archive, when summoned, reanimate what has been, nearly, forgotten.

But the archive as a place that holds histories waiting to be unearthed is also constantly at risk in its potential excess of being wholly unknowable or beyond analysis (and thus practically invisible). Assmann writes that "the storage capacity of the archive has by far exceeded that which can be translated back into active human memory";[28] for all its will to make knowable, the archive thus risks becoming impenetrable in its conglomeration of information. In this volume, both Camilla Salvaneschi and María A. Cabrera Arús, underscore a simultaneous need for the expansive archive and a grappling with the question of how to make sense out of what is accrued. Their work underscores a proactive, self-archival turn aided by digitization and online dissemination. The inception of the digital archive at once potentially exacerbates an overload of data or gives a false sense of total representation, while also circumventing material concerns with regards to storage capacities, and offers significant opportunity for the democratization of access and analytical modes of processing, so that even an abundance of archived items do not "fall out of the frames of attention, valuation, and use."[29]

Although a wide range of institutions, and the archivists and librarians who work there, have been essential in providing access to the materials that are engaged in this book, the recently generated collections discussed here are typically those undertaken by a small network or community that responds to gaps in preexisting archives. For it turns out that for all its will to "collect everything,"[30] many things have indeed been left out of the archive. Archives are not a mere random conglomeration of materials – they do represent a coherence to a specifically outlined collection and the willful curation and cataloguing of the archivist, the outcome of which is far too often that, at least in the West, the archives contain a lot (one might say an excess, even) of materials pertaining to a certain kind of history and a relative paucity of items pertaining to anyone or anything else.[31] If the objects are not there in the archive to be summoned in the first place, it becomes that much more difficult to resurrect and reevaluate that history, to move it from "past past" to "past present." Therefore, there is a real need for constructing alternative archives and institutions, while at the same time reorganizing and improving the dominant ones, laying bare the gaps that arise, for instance, from what Eric Gardner describes as "racist collection-development policies" in reference to both physical and digital holdings, fraught with long histories of bias at the basic level of collecting and cataloguing.[32]

In 2016 at the Radcliffe Workshop on Technology and Archival Processing, the archivist and anthropologist Jarrett Drake addressed these not-accidental gaps in a succinct critique of the archival principle of provenance.[33] At the conclusion of his talk Drake asked that if provenance has worked to make most visible in the archive "white wealthy Western men":

> how might one interpret these silences and absences [of materials pertaining to other identities] not as sudden surprises but as predictable products of archival processes and principles rooted in colonialism, anti-black racism, sexism, homophobism, transphobism, and classism? Most importantly, how can archivists revisit this core principle to learn of its limitations and envision a post-colonial archive free of these oppressive forces and equipped to meet the challenges of contemporary born-digital archival records?[34]

An exhibition curated by Tímea Junghaus at the Budapest Gallery8-Roma Contemporary Art Space in April 2015 titled *Archive of Desire* offers an evocative and creative response to this question, and is another instance of a contemporary practice of intervention in reclaiming the archive, that also engages print materials. Junghaus looks to collections representing Roma populations – specifically in photographs – held in museums in Central Europe, and argues that,

> From the captions that are attached to these archival materials, it is clear that the Roma cultural heritage, the Roma past, and pictures of our Roma ancestors are lying in archives that register only the names of the collectors and ethnographers, not the people in the photographs.[35]

In response, images found in these collections – which, according to Junghaus include, "an outrageous number of pictures of Roma that either serve the rather prurient desires of the collectors or are simply indecent and offensive" – were displayed in her exhibition "in miniaturized form: visitors could view them only under magnification. This subversive curatorial strategy was a gesture that emphasized the surveillance and voyeurism. The magnifiers not only enlarged the images, but also stressed the significance and nature of the gaze."[36]

In reclaiming and recontextualizing the images of those with whom she has a shared past, Junghaus performs a strategy that is an evocative potentiality for engagement with old archives in new times. Other approaches to archiving a personal history as it relates to a larger legacy are presented in several of the chapters here, as in the Israeli "one-time periodicals" discussed by Nana Ariel that illustrate a visceral grappling with a troubled history of multinational politics and occupation while reflecting the very individual experience of forced migration,

Jennifer Bajorek's discussion of how the artist Kelani Abass employs old family photographs as a way to layer multiple histories and print technologies, or in Cabrera Arús's personal intervention in preserving and disseminating state records through the maintenance of a digital archive platform.

The digital might offer a particularly salient tool for active participation and intervention into processes of collecting and cataloguing. But, as Gardner warns, with the development of "digital humanities initiatives, we cannot afford to let budgets for special collections continue to drop. We should never consider digital and paper an 'either/or' condition."[37] With the digitization, and thereby activation, of an archive by a wider body of viewers and researchers, there must be a concurrent impulse within archives, libraries, and museums to facilitate access to a wider range of material objects for more diverse publics, and concurrently, greater consideration on the part of researchers for the labor that goes into creating and maintaining these collections. In *Spontaneous Particulars: The Telepathy of the Archives*, Susan Howe offers, in the wake of the radical transformation brought about by the digital, a "swan song to the old ways."[38] She writes eloquently of the materials waiting to be summoned, their pastness resurrected for the present:

> In research libraries and collections, we may capture the portrait of history in so-called insignificant visual and verbal textualities and textiles. In material details. In twill fabrics, bead-work pieces, pricked patterns, four-ringed knots, tiny spangles, sharp-toothed stencil wheels; in quotations, thought-fragments, rhymes, syllables, anagrams, graphemes, endangered phonemes, in soils and cross-outs.[39]

The impossible goal but necessary task of the archive to contain "everything" cannot reside only at the level of the digital. For anyone seeking its mysteries should have the hope of entering the physical space of the reading room and finding the "thought-fragments" that speak to them, having access to the haptic pleasure of connecting with its treasures, or reckoning up close with its hard truths.

Image, object, text

The chapters in this volume are not organized chronologically or geographically, but rather broadly across the categories of "Image," "Object," and "Text." This is an oblique reference to the English-language title of a translated volume of collected essays by Roland Barthes, which were originally written in the 1960s and 1970s: *Image, Music, Text*. Barthes's semiotics across these essays evinces a deeply material approach to the three elements of the title, as he considers for example the nature of the press photograph, the transmission of sound, the very presence of the author. Similarly, earlier essays, from 1954 to 1956, collected in *Mythologies*, together come to represent a sort of archive, or a cataloguing of an archive, that collects and analyzes some objects drawn from "myths of French daily life" as encountered in "(a newspaper article, a photograph in a weekly, a film, a show, an exhibition)."[40] Detergent, toys, margarine, milk, and a photograph are all materials in this Barthesian archive. The present volume also offers its own archive of objects that participate in the cultural histories that are being told here: it contains the rubber stamp and the printing press, photographs and mimeograph machines, current converters and radios, ID booklets and printed posters.

An engagement with Barthes lays bare the Western literary and art historical education I have inherited, and offers to my mind a compelling way to organize the chapters here and frame the three distinct sections. But the texts to come are not themselves explicitly (or not at all) in dialogue with this legacy. And if it will seem to the reader that a chapter that appears under the subsection of "Object" could just as well relate to "Image," or an entry under "Text" to either of the other two categories, that is in fact part of the point: the nine chapters included here have been conceived to stand together in a way that erodes not only canonical conceptions of Western, modernist print narratives, but the very disciplinary boundaries that are often constructed between fields of discourse that prevent more non-hierarchical readings of image and text-based works. The reader is encouraged to read the individual chapters against the very word under which they have been categorized, and consider what is productive in thinking about their given content within that particular subject heading, and outside of it.

Image

Since the inception of photography, the medium has had a marked impact on theories of vision, other art forms, such as painting, and indeed the role of the archive.[41] The almost oppressive omnipresence of photographic images in our daily life today, thanks to the cameras on our mobile devices and the 24-hour news cycle, secures photography's status as the image maker *par excellence*. Barthes considers the presumed indexicality of the photograph – its presentation of "the stupefying evidence of *this is how it was*"[42] – and its simultaneously elusive capacity – "a decisive mutation of informational economies."[43] He is a central figure to those studying and writing the history of photography in the West, which is a history of photographic theory and photomechanical discourse rooted in a broader context of capitalist exchange, where the advertising image, for instance, offers a ready and oft taken up example. But the three chapters in this first section engage histories of print production that push against the taken-for-grantedness of this history of image production or reveal how modes of advertising and consumerism operate outside of that context.

Jennifer Bajorek, in the chapter that opens this volume, "Analog and digital Nigeria: inheriting serial cultures in the work of Kelani Abass," asks,

> What if we acknowledged this coexistence [of manual and mechanical reproduction processes in West Africa] more fully than it has been in the scholarship to date? Would this not alter the dominant paradigms for thinking about both print and photography – in West Africa and in the rest of the world? (27)

In a case study of the work of Abass, Bajorek brings to the fore a consideration of print histories that incorporates not only the photographic medium, but also that of the family commercial printing studio, a site where Abass spent countless hours of his early life. By considering how the legacy of a family business is reflected in an art practice that overlays references to various print discourses and histories, Bajorek resists a privileging of technological innovations sheerly for their "'mechanical reproducibility,' understood narrowly within the logics of commodity production and of mass production dictated by industrial capital (26)." Abass's work is framed as a mode of archiving at the local level, a representation of family history and inheritance, that also, through the employment of overpainting and collage, is part of "West African photography histories that have veered sharply away from a concern with photographic indexicality [. . .] to underscore the relentless mutability of the photographic object (30–31)."

Moving from photography to film, Olympia Bhatt, in her chapter, "Writing about sound: the early talkie film periodicals of India," looks at how aurality was imagined in popular Indian film magazines. She shows how the emergence of talkies in the late 1920s was figured in advertisements, with reference to the recording technologies used and the valorization of stars, often pictured in movie stills. Some of the advertisements that Bhatt presents were meant to summon the intrigue and grandeur of the talkies in the "picturization" of sound. One magazine cover from *Filmindia* alluringly visualizes the presence of sound in film, as two female stars are illustrated riding on bicycles (a popular way of picturing the "new woman" in the West as well) and singing in Hindi, the very wheels of their cycles turned into gramophone records, thus serving as musical accompaniment as they ride along. These illustrated film magazines, "within the rapidly evolving soundscape of early twentieth-century India," thus document the country's shifting relationship towards sound in film while also participating in "the industrial infrastructure created to supplement the commodity culture around cinema (46)."

Nana Ariel, in her discussion of Hebrew "nonperiodical" periodicals, looks to a decidedly noncommercial mode of publication, and contests a presumption of the "little magazine" – especially one with only a single issue – as "failure," for all its lack of capitalist potential. Her chapter, "One-time periodicals, unrealized manifestos: little magazines and the ethos of failure," opens with a discussion of a portrait of the crew of the ill-fated space shuttle Columbia included in the magazine *02010* (published in 2010), and suggests that this loaded image here is meant to reference and subvert the sort of "(magnificent) failure (82)" that alternative magazines themselves embody. Ariel contends that the image of publications such as *02010* as failures is in fact a self-conscious, ironic performance of their makers, and that these failures, for their charged capacity as salient cultural markers, are, in fact, successes. Nevertheless, in the case of the magazine *Kiltartan*, produced as a single issue in 1964 in a run of 500 copies, it is almost entirely outside the realm of visibility today. Ariel writes that, "Only a few of these copies, circulating among random collectors and auction houses, remain, and to date *Kiltartan* is not included in any official collection or archive, with only one copy in the Israeli National Library (90)." And yet, the very object of *Kiltartan*, which is a boxlike structure containing items of various sizes, is a material monument (an archive) to a specific moment in the early days of the formation of the state of Israel that successfully preserves a rare example of a counter narrative to the mainstream political discourse, permitted to exist precisely because it was presumed to be so marginal.

Object

The texts and images that assemble to comprise *02010* or *Kiltartan* might well also be described for their objectness, and while these publications themselves are at a remove from the mainstream, their contents, such as the inclusion of the Columbia crew photo in *02010*, do certainly reference popular culture, which so often comes under Barthes's scrutiny. In *Mythologies*, he looks to those objects that have undergone an "ideological abuse" so that with a certain "naturalness" the "newspapers, art and common sense constantly dress up a reality which, even though it is the one we live in, is undoubtedly determined by history."[44] It is thus worth questioning, the "what-goes-without-saying" valence of the objects of our everyday, and reading them within their historical context so that we can better understand those histories and contexts.[45] Barthes performs a powerful unpacking of the seemingly innocuous child's plaything, about which he writes,

> French toys: one could not find a better illustration of the fact that the adult Frenchman sees the child as another self. [. . .] French toys always *mean something*, and this something is always entirely socialized, constituted by the myths or the techniques of modern adult life: the Army, Broadcasting, the Post Office [. . .].[46]

If the objects say something about their maker, and likewise reflect the social apparatus within which they were constructed and exchanged, then careful attention to these material goods – be it obscure avant-garde magazines or plastic tchotchkes of popular consumption – and their circulation and preservation open up a wider conversation on the political structures within which they exist, which the following three chapters pursue.

In "Via postal: networked publications in and out of Latin America," Zanna Gilbert offers a reading of a set of mail art magazines that is deeply embedded in a consideration of their materiality. The mail art "assembling" becomes the site at which items are collected, and also the vessel by which those same items are disseminated, a sort of moving and multiple archive. Gilbert writes that, "The magazines were key in constructing the translocal communication engendered by the network, which in turn enabled the formation of a conceptual community between scattered groups of people who coalesced around common aesthetic interests and worldviews (105)." If the mail art networks facilitated local transmission through the process of assembling, they also, in their broad dissemination via the post – that iconic location of governance, socialization, and circulation – were capable of "distributing uncensored information across vast distances, [and] as such they provide an early account of an incipient globalization of art networks (127)." The case of the mail art magazine(s) offers a mode by which discussions occurring on the very local level can readily resonate on a more broadly international scale, in this case through distribution methods in print and post.

And today, the digital marks the capacity to bring the hyperlocal to the level of the global. María A. Cabrera Arús, in her chapter, "The life of others: collecting and archiving the Cuban surveillance regime," describes how a digital archive she started in 2012 of "appliances, furniture, clothing, ephemera, and memorabilia" represents "the largest repository of material culture of Cuban socialism" (which is freely accessible online) (134). The collected items either originating in, or imported to, Cuba from the Soviet Union in the period from the 1960s to 1980s themselves embody a network between socialist Cuba and socialist Eastern Europe that was eroded with the collapse of Communism in the latter geography around 1989. Cabrera Arús suggests that this particular network in history risks being forgotten, but through an online archive of her own making she seeks to provide a site that helps facilitate research and collective memory. Cabrera Arús circumvents the "state archives [which] are closed to public scrutiny, (145)" and here, she considers Cuba Material's collection of printed ID cards and other bureaucratic documents that suggest the "egalitarian, nationalist, and modernizing narratives that on occasion connected the postrevolutionary present with the prerevolutionary past or with both the Soviet Bloc and the Third World imaginaries (134–135)." The online archive, which makes publically visible this pan-socialist network, now also activates a community of Cubans who donate items to the collection, facilitated through social media networks such as Facebook.

The physical object of the printed magazine, and the online archive that is often a container for its digital second self, is the focus of Camilla Salvaneschi's chapter, "Contemporary art magazines: the archive in the archive." Salvaneschi, through a consideration of contemporary strategies enacted both by art magazine editors and organizations that maintain online libraries of previously printed materials, "propose[s] a reading of the magazine deeply linked to the archive, as both are characterized by complex temporalities, and both the magazine and the archive, have been considered historical repositories of documents (158)." By looking at a variety of cases across a broad geography – from Brooklyn to Hong Kong, Cape Town to Zurich – Salvaneschi underscores the modes by which the magazine can function both as a physical and digital platform for a wider exchange, and asks "How does the magazine use the archive as a site of network formation, to expand its own geographical boundaries? (158)" By also considering the art magazine as an archive itself, she suggests its empowering potential as a platform that imag(in)es diverse art histories and contemporary art narratives. As the magazine becomes an object within the digital archive, those perspectives contained within it are thus propagated further.

Text

The printed or digital object of the magazine, in Salvaneschi's conception, is a receptacle of storage for both image and text. The volume's final section begins with close readings of those texts that comprise a publication, and moves to consider the very script that forms an alphabet. In his essay "From Work to Text," Barthes argues that "against the traditional notion of *work* [. . .] there is now the requirement of a new object, obtained by the sliding or overturning of former categories. That object is the *Text*."[47] Moving now from a cluster of chapters ostensibly grouped under the category of "Object" to those corresponding to "Text," the latter is taken on here in a way that gestures towards what Barthes might have had mind, when he writes of the text standing at an "intersection" of "propositions," which include: "method, genres, signs, plurality, filiation, reading and pleasure."[48] The material of the text can call for formal readings – or consumption – of the very marks that comprise a word or sentence as a work of art, and the boundaries between text and image thus become blurred. In the final three chapters, close considerations of the nature of a text – how it came into being and to whom it is meant to speak – invites an engagement with broader issues of the social and political import of the printed word.

In his chapter, "A journal of science without boundaries: Ali Suavi's *Ulûm Gazetesi*," Kenan Tekin offers close textual readings of a short-lived late Ottoman publication produced by one man in Paris. Tekin describes how Ali Suavi, a member of the Young Ottomans, took advantage of greater "access to printing technologies [that] expanded in the nineteenth century," and a world that had become "increasingly interconnected through faster modes of travel and communication, [allowing] intellectuals and journalists who were barred from expressing their views [at home to] seek alternative dominions whence they would publish freely (193)." *Ulûm Gazetesi*, though in print for only a year, from 1869 to 1870, became a salient platform through which Suavi could offer informative articles on the sciences while also presenting his opinions on various social, political, and religious issues. Living in London and Paris, Suavi also witnessed firsthand European imperialist and racist views, and used his platform to counter these. In a questionnaire published in *Ulûm Gazetesi*, for instance, Suavi engages a fictionalized interlocutor in debates around whether the Ottoman Turkish alphabet should move away from Arabic script (which would come to pass only in the early twentieth century), to which Suavi ultimately takes an ambivalent stance, unwilling to embrace a full turn towards Western systems of writing.

Hannah Pröbsting recounts a comparable debate – in her chapter "Everyday printed matter: Kurt Schwitters's experimental typography" – erupting in Germany in the late nineteenth century over "which kind of script – the ornate *Fraktur* (Gothic) or the simpler *Antiqua* (Latin) – was most suitable for texts published in German (200)." But in the 1920s, when a group of avant-garde practitioners moved definitively towards the more functionalist (and Latin) style of New Typography, one of its main proponents, Kurt Schwitters, began to envision a wholly new system of writing that would transcend the local German context, which he theorized and illustrated in a magazine article in 1927. As Pröbsting relates, "What Schwitters envisaged was not just a European version of internationalism, but a much more radical one, bringing together sounds from all world languages in his *Systemschrift* (209)." Although the script, cryptic and arcane, did not make its way definitively into wide usage, Pröbsting points to a few appearances of a simplified version of the text that did appear in the public sphere, on posters and signboards, and considers the material and digital legacy of the letterforms today.

While Schwitters emphatically aimed for a practical legibility for his script as language, Naomi Kuromiya outlines formalist debates around the consideration of calligraphy not only as text, but image, in "The *sekai-sei* of circulation: the 'world relevance' of the avant-garde Japanese calligraphy periodical *Bokubi* (1951–1960)" Kuromiya describes how *Bokubi* reproduced not only examples of "modern" Japanese calligraphy, but also works by contemporary Western artists working in a so-called calligraphic style, a strategy employed by its editor Morita Shiryū for gaining greater visibility on a world stage. Taking an example from an exhibition of Japanese calligraphy at The Museum of Modern Art in New York in 1954, Kuromiya observes that "only when calligraphy was freed from its traditional, lexical qualities, could it serve as inspiration for 'original' modernist abstract painting as canonized by MoMA (237)." The ostensibly positive outcome of finding greater international recognition for the work of the Japanese calligraphers represented in *Bokubi* points also to the problematic condition that in order for the Japanese calligraphic tradition to be "legible" in a more global context it had to first be read within Western art historical narratives, essentially stripping it of its textual function. Nevertheless, in the post–World War II era the magazine served as a vehicle for what might seem an unlikely network formation and transnational artistic exchange between Japan and the US and France.

Kuromiya's chapter brings us full circle to where we began, with a consideration of the image. And it is here that this introduction ends and the chapters discussed above begin. Barthes writes in *Image, Music, Text* that, "The metaphor of the Text is that of the *network*."[49] It is the work of this volume to suggest a wide range of networks, within the subjects covered in the individual chapters, and across the full body of the book, in ways that might not have been previously considered were not these disparate materials placed side by side here.

International Perspectives on Publishing Platforms as a whole, and each chapter individually, reflects an effort to alter the field of vision within periodical studies and related disciplines. This methodology will no doubt also invite the reader to see what is not here. There were other histories that I had hoped to include, and far more I can be sure that I will have failed to have thought of, but other readers will no doubt find to be lacking. The discursive framework within which this book was envisioned is nevertheless precisely meant to be able to open itself up to, and invite, such a critique. These nine chapters are naturally not intended to be representative of the full scope of periodical and archival studies at this digital moment. It is rather my hope that the wide range of examples presented here, each representing a detailed case study within a specific subject area, will together offer but one engagement of many to come that helps to chart a more dynamic field.

Notes

1 I am grateful to Vlad Beronja and Hannah Pröbsting for their helpful comments on an earlier version of this introduction. I also wish to thank the Institute for

Public Knowledge at New York University – in particular Siera Dissmore and Gordon Douglas – for hosting the conference "Summoning the Archive: A Symposium on the Periodical, Printed Matter, and Digital Archiving" that I organized in May 2017, and which provided an early platform for the conversations that this volume takes up.

2 Indeed, recent publications initiated by archivists and librarians seek to address and offer solutions to these issues. See, for instance, Melissa Morrone, ed., *Informed Agitation: Library and Information Skills in Social Justice Movements and Beyond* (Sacramento, CA: Library Juice Press, 2014); K. R. Roberto, ed., *Radical Cataloging: Essays at the Front* (Jefferson, NC: McFarland & Co., 2008).

The librarian Sanford Berman has been advocating for decades for more critical interrogation of accepted collection and cataloguing practices. With regards to Library of Congress subject headings, he has written that while they are "a real boon to scholars, as well as to ordinary readers," they are also insufficient in terms of a flexibility of representation, as they can "only 'satisfy' parochial, jingoistic Europeans and North Americans, white-hued, at least nominally Christian (preferably Protestant) in faith, comfortably situated in the middle- and higher-income brackets, largely domiciled in suburbia, fundamentally loyal to the Established Order, and heavily imbued with the transcendent, incomparable glory of Western Civilization." [Sanford Berman, *Prejudices and Antipathies: A Tract on the LC Subject Heads Concerning People* (Metuchen, NJ: Scarecrow Press, 1971), ix.] My thanks to Jennifer Tobias for bringing the work of Sanford Berman to my attention.

Some forty years later, in *From Polders to Postmodernism: A Concise History of Archival Theory*, John Ridener describes shifts in the "contemporary archival paradigm" brought about with cultural change, for instance with the "Civil Rights, Women's Rights, Gay Rights, and anti-war movements of the twentieth century, coupled with the entrenchment of the Cold War and possibility of total destruction through nuclear war. [. . .] Archivists who established the contemporary archival paradigm continue to change and shape theoretical and practical answers to the new needs of archive users and their professional peers." [John Ridener, *From Polders to Postmodernism: A Concise History of Archival Theory* (Duluth, MN: Litwin Books, 2009), 102.]

3 See, for instance: Suzanne W. Churchill and Adam McKible, eds., *Little Magazines & Modernism: New Approaches* (Aldershot: Ashgate, 2007); Jayne E. Marek, *Women Editing Modernism: "Little" Magazines & Literary History* (Lexington: The University Press of Kentucky, 2015); Mark S. Morrisson, *The Public Face of Modernism: Little Magazines, Audiences, and Reception, 1905–1920* (Madison: University of Wisconsin Press, 2001); Eric B. White, *Transatlantic Avant-Gardes: Little Magazines and Localist Modernism* (Edinburgh: Edinburgh University Press, 2013); Robert Scholes and Clifford Wulfman, *Modernism in the Magazines: An Introduction* (New Haven, CT: Yale University Press, 2010).

While I argue that early twentieth-century Western, white modernism and the "little magazines" take up the dominant discourse in the field, there are of course other existing books related to periodical studies that reach beyond this canon. To name just a few, Géza Perneczky's *The Magazine Network: The Trends of Alternative Art in the Light of Their Periodicals, 1968–1988* (Köln: Soft Geometry, 1993) is the English translation of a volume originally published in Hungarian with support of the Soros Foundation and deals extensively with Eastern European and Latin American art represented alongside North American trends from the period; the recent re-edition of *Wimmen's Comix* in 2016, with an introduction by Trina Robbins, brings renewed attention to this groundbreaking feminist serial publication of the 1970s; and Noliwe Rooks's, *Ladies' Pages: African American Women's Magazines and the Culture That Made Them* (New Brunswick, NJ: Rutgers University Press, 2004) takes up issues of gender and sexuality within the context of African American migration and urbanization through the lens of the magazine.

4 On the point of the periodical as archive, Jacques Derrida, in his well-known *Archive Fever*, goes so far as to name the "press, printing, ink, paper" as the "archiving

machine" in a discussion of Freud's writings, and then adds "portable tape recorders, computers, printers, faxes, televisions, teleconferences, and above all E-mail." Derrida's consideration of the archive thus does not concern so much the architectural site of the archive as *that which is archived*, as well as the media of production/dissemination. [Jacques Derrida, *Archive Fever: A Freudian Impression* (Chicago: University of Chicago Press, 1996), 8, 16].

Further, in her recent book *Archive Everything: Mapping the Everyday*, Gabriella Giannachi outlines a trajectory for archives that "over the centuries" have come "to be considered not only as locations or objects but, increasingly, as media, and communication strategies." Gabriella Giannachi, *Archive Everything: Mapping the Everyday* (Cambridge, MA: MIT Press, 2016), 9. A heterogeneous and multimedial approach to the archive is likewise taken up here.

5 In "Theories of the Archive from Across Disciplines," former MIT librarian Marlene Manoff describes the work of Paul Otlet in the 1930s, a founder of the documentation movement, to suggest a definition of archival documents as "simply objects that conveyed information and thus the term could refer to anything collected by archives, museums, or libraries. As libraries, museums, and archives increasingly make their materials available online in formats that include sounds, images, and multimedia, as well as text, it no longer makes sense to distinguish them on the basis of the objects they collect." Marlene Manoff, "Theories of the Archive from Across the Disciplines," *Portal: Libraries and the Academy* 4, no. 1 (January 2004): 10. This object-oriented view towards the archive, and related library and museum collections, resonates across the chapters in this book and will be discussed in further detail in the final section of this introduction.

6 In the essay "Exhibition as Archive," Beatriz Colomina offers a succinct definition of the "little magazine": "an Anglo-Saxon term first used to describe small avant-garde literary publications, such as Margaret Anderson's *Little Review* of the 1910s and 20s, that were dedicated to progressive theory, art, and culture. They were set apart from established periodicals by their noncommercial operations and small circulation." [Beatriz Colomina, "Exhibition as Archive," in *The Archive as a Productive Space of Conflict*, ed. Markus Miessen and Yann Chateigné (Berlin: Sternberg Press, 2016), 168]. As Colomina describes them, the "little magazines" themselves were generative sites of agitation and change on the sociopolitical level that ought to be "analyzed as systems" and the term magazine "not taken at face value," so as to also include related ephemera such as postcards, posters, and advertisements (169). This broader conception invites the recognition that the rhetoric of resistance and capitalist circumvention for which the prototypes of the "little magazine" are defined is not their exclusive domain, though they have received far more attention than other radical materials that fall outside of that "Anglo-Saxon" frame of the early twentieth century.

7 Johanna Drucker, *The Visible Word: Experimental Typography and Modern Art, 1909–1923* (Chicago: University of Chicago Press, 1994), 81.

8 Scholes and Wulfman, *Modernism in the Magazines*, 27.

9 Karel Teige, "Umění dnes a zítra," in *Revoluční sborník Devětsil* (Prague: Večernice V. Vortel, 1922), 193. Emphasis is the author's. Translation is my own.

10 Johanna Drucker, "Le Petit Journal des Refusées: A Graphical Reading," *Victorian Poetry* 48, no. 1 (2010): 147.

11 Several volumes have attempted to address this issue directly. See, for instance, Sanja Bahun-Radunović and Marinos Pourgouris, *The Avant-Garde and the Margin: New Territories of Modernism* (Newcastle, UK: Cambridge Scholars Press, 2006); Chana Kronfeld, *On the Margins of Modernism: Decentering Literary Dynamics* (Berkeley: University of California Press, 1996); Partha Mitter, *The Triumph of Modernism: India's Artists and the Avant-Garde, 1922–1947* (London: Reaktion Books, 2007); Elaine O'Brien, Everlyn Nicodemus, Melissa Chiu, Benjamin Genocchio, Mary K.

Coffey, and Roberto Tejada, eds., *Modern Art in Africa, Asia, and Latin America: An Introduction to Global Modernisms* (Chichester, West Sussex and Malden, MA: Wiley-Blackwell, 2012).

12 Robin Kinross, *Modern Typography: An Essay in Critical History* (London: Hyphen Press, 1992), 9.

13 A useful concept here is the term "uneven modernity," which Harsha Ram employs, for instance, in his discussion of Georgian modernism, to describe "the condition in which cultural forms, socio-economic structures and ideological projects associated elsewhere with distinct historical periods or modes of production coexist in close proximity." [Harsha Ram, "Introducing Georgian Modernism," *Modernism/Modernity* 21, no. 1 (January 2014): 285.]

14 Kronfeld, *On the Margins of Modernism*, 5.

Thomas S. Mullaney's new book, *The Chinese Typewriter: A History*, speaks directly to the Latin-alphabet bias in the West as it relates to global modernism; he describes in a chapter titled "Incompatible with Modernity" a concerted effort by typewriter manufacturers such as Olivetti and Remington to develop keyboards with symbols that could accommodate a variety of languages (including, for instance, Arabic, Hebrew, Russian, and Khmer), thereby reflecting "a new global modernity." [Thomas S. Mullaney, *The Chinese Typewriter: A History* (Cambridge, MA: MIT Press, 2017), 58.] Mullaney describes how an engagement by these firms with various writing systems, with "the taken-for-grantedness of English" constituting the keyboard against which all others were adapted, meant that "the alphabetic and syllabic scripts of the world were ranked on a scale of greater and lesser compatibility with the modern" (60). The failure to find a solution for a single-shift keyboard typewriter for Chinese provided fodder in the West for depictions of "Chinese antimodernity" and "linguistic unfitness" (70).

15 Kinross, *Modern Typography*, 10.

16 "Princeton Blue Mountain Collection," accessed October 25, 2018, http://bluemountain.princeton.edu/bluemtn/cgi-bin/bluemtn.

17 "Modernist Journals Project," accessed April 1, 2018, http://modjourn.org/about.html.

18 "The International Dada Archive," accessed April 1, 2018, http://dada.lib.uiowa.edu/.

19 "Monoskop," accessed April 1, 2018, https://monoskop.org/Monoskop.

20 Asia Art Archive, "Home," accessed April 1, 2018, https://aaa.org.hk/en.

21 "Chimurenga: A Pan African Publication of Writing, Art and Politics," *Chimurenga*, accessed April 28, 2018, www.chimurenga.co.za/.

Both the Asia Art Archive and Chimurenga will be discussed in Camilla Salvaneschi's chapter in this volume.

22 As a counterexample, I myself maintain a micropublishing platform that has been steadfastly committed since its inception in 2011 to *not* providing digital access to its content. "Avant-garde in retrograde," harlequin creature employs so-called "obsolete technologies," such as the letterpress and Risograph machine, as well as the typewriter. Edition sizes are limited to 100 copies. What might be described as the radical inaccessibility of this project is meant not to create a precious object for wealthy patrons but rather to ask the person who engages with these material forms to reflect on the haptic, experiential qualities exclusive to printed matter, and to thereby highlight the importance of slow work and offline community building within the digitally driven society in which we live.

23 In "Implications of Archival Labor," notes from a panel discussion at the 2016 Organization of American Historians, Stacie Williams discusses the invisibility of the work of the archivist, and need for advocacy around issues related to salary and better understanding and respect for the labor of workers in this field. She writes of the disconnect between the research that a scholar does, and the scholar's understanding of how those materials to which they have access have come to be made available: "They are hungry for research or information in our collections, but very little thought goes into the team of people who make it possible: the collections management archivist, the manuscript

archivist, the technical services cataloger, the digital archivist, the reference archivist, and most importantly, the people who actually process the collections." But she also underlines the complementary necessity of self-reflection and fundamental change within the field itself: "Perhaps we are so terrible at advocating for the importance of what we do because to be good at that advocacy means acknowledging that the manner in which we conduct this labor is often times unequal, rooted historically in sexism, racism, ableism, and classism, and that will always present a challenge to the access we hope to provide." [Stacie Williams, "Implications of Archival Labor," *Medium* (blog), April 11, 2016, https://medium.com/on-archivy/implications-of-archival-labor-b606d8d02014. Last accessed July 23, 2018.]

24 Birte Kleeman, "I Put This Moment Here," in *Wunderkammer*, ed. Birta Kleeman (Brooklyn: Brooklyn Fine Arts, 2012), 8.

25 Aleida Assmann, "Canon and Archive," in *Cultural Memory Studies: An International and Interdisciplinary Handbook*, eds. Astrid Erll and Ansgar Nünning (Berlin: Walter de Gruyter, 2010), 98.

26 In *The Big Archive*, Sven Spieker writes of this process in relation to archives initiated as a bureaucratic necessity: "as they enter the archive, the papers of which offices rid themselves are resurrected as sources that historians consult in their efforts to write history. From the historian's point of view these papers stand as quasi-objective correlatives of the living past." [Sven Spieker, *The Big Archive: Art from Bureaucracy* (Cambridge, MA: MIT Press, 2008), xii.]

Similarly, but with a rather more critical view, Paul Voss and Marta Werner write in their introduction to a special issue of *Studies in the Literary Imagination* the paradox by which the archive "preserves and reserves, protects and patrols, regulates and represses. The architecture of the archive and the sentinels who control access to its interior suggest that the conservation and transmission of knowledge has been, at least historically, the prerogative of a few chosen agents, of a coterie of privileged insiders. As scholars, we benefit from this privilege." [Paul J. Voss and Marta L. Werner, "Toward a Poetics of the Archive: Introduction," *Studies in the Literary Imagination; Atlanta, Ga.* 32, no. 1 (Spring 1999): i.]

27 Assmann, "Canon and Archive," 103.

28 Ibid., 104.

29 Ibid., 97–98.

While the state archive is largely outside the scope of this introduction, it should be noted that besides instances in which archives fail to collect certain materials, in the case of what Assmann describes as "political archives" – those that "function as an important tool for power" – there is often highly regulated access to the materials held within, thus actively keeping those histories dormant (102).

30 Spieker writes that, "where archives have to collect everything, because everything may become useful in the future, their storage capacities are soon exhausted" (22).

31 There are also examples of archivists who make major, proactive efforts to ensure representation of specific communities in an archive. Shawn(ta) Smith, for instance, works tirelessly to ensure that the histories of black lesbian women are preserved in the holdings of the Lesbian Herstory Archives in Brooklyn, New York, by personally walking women through the process of donating items to the collection. She has created a zine that showcases materials in the collection related to Black Lesbian Herstory, alongside information from the archive's website with regards to its aims, accessibility, and the necessary steps for creating a special collection out of "letters, emails, birthday cards, photographs of friends, lovers, and family, your unpublished poetry and prose." The zine then includes a page of numbered lines, titled "My Special Collections List," encouraging women to fill it in with "Stuff I'm Gonna Donate to the Archives . . ." [Shawnta Smith, *Black Lesbians in the 70's and Before: An at Home Tour at the Lesbian Herstory Archives* (Brooklyn, NY: Lesbian Herstory Archives, 2010.)] For more on the archive, see Shawn(ta) Smith-Cruz, Flavia Rando, Rachel

Corbman, Deborah Edel, Morgan Gwenwald, Joan Nestle, and Polly Thistlethwaite, "Getting from Then to Now: Sustaining the Lesbian Herstory Archives as a Lesbian Organization," *Journal of Lesbian Studies* 20, no. 2 (April 2, 2016): 213–233; and visit "The Lesbian Herstory Archives," accessed May 5, 2018, www.lesbianherstoryarchives.org.

32 Eric Gardner, "Accessing Early Black Print," *Legacy: A Journal of American Women Writers* 33, no. 1 (2016): 26.

33 Jarrett M. Drake, "RadTech Meets RadArch: Towards A New Principle for Archives and Archival Description," *Medium* (blog), April 6, 2016, https://medium.com/on-archivy/radtech-meets-radarch-towards-a-new-principle-for-archives-and-archival-description-568f133e4325#.86ba9wqwf. Last accessed July 23, 2018.

While there is no room here for a more detailed history of the development of provenance, in *The Big Archive*, Spieker provides the German context for the term: "In 1881, the so-called *Provenienprinzip* or principle of provenance (PP) was introduced at the Privy State Archive in Berlin. It stipulated that archival files were to be arranged in strict accordance with the order in which they had accumulated in the place where they had originated *before* being transferred to the archive" (17).

And Giannachi, in *Archive Everything*, offers an outline of the principles of provenance as introduced in the *Manual for the Arrangement and Description of Archives*, originally published in Dutch in 1898 by Samuel Muller, Johan Feith, and Robert Fruin: "This seminal work articulated the principles concerning the nature and treatment of archives, including the fact that archives from different creators must not be mixed or based into artificial arrangements dependent on chronology, geography, or subject, but rather that the arrangement must be based on the original organization of the collection, which may in turn reflect the administrative body that produced it" (6–7).

34 Drake, "RadTech Meets RadArch."

35 Tímea Junghaus, "Our Beloved Margins: The Imaginings of the Roma Transformative Subject and Art History Scholarship in Central Europe," in *Extending the Dialogue: Essays by Igor Zabel Award Laureates, Grant Recipients, and Jury Members, 2008–2014*, ed. Urška Jurman, Christiane Erharter, and Rawley Grau (Ljubljana: Igor Zabel Association for Culture and Theory, 2016), 86.

36 Ibid., 84.

37 Gardner, "Accessing Early Black Print," 26.

38 Susan Howe, *Spontaneous Particulars: The Telepathy of Archives* (New York: New Directions, 2014), 9.

39 Ibid., 21.

40 Roland Barthes, *Mythologies*, trans. Annette Lavers (London: Grant and Cutler, 1994), 11.

41 Citing the writings of Hal Foster ["An Archival Impulse" *October* 110 (2004): 3–22] and Okwui Enwezor [*Archive Fever: Uses of the Document in Contemporary Art* (New York, NY: International Center of Photography; Göttingen: Steidl Publishers, 2008)], Gabriella Giannachi describes "technological advances" that can be attributed to a "proliferation of archives" in the twentieth century that "emerged in the aftermath of the invention of photography" (*Archive Everything* 14).

42 Roland Barthes, *Image, Music, Text*, trans. Stephen Heath (New York: Noonday Press, 1978), 44.

43 Ibid., 45. For a sustained later engagement by Barthes with photography and its indexicality, see *Camera Lucida: Reflections on Photography*, trans. Richard Howard (New York: Hill and Wang, 1981).

44 Barthes, *Mythologies*, 11.

45 Ibid.

46 Ibid., 53.

47 Barthes, *Image, Music, Text*, 156.

48 Ibid.

49 Ibid., 161.

Bibliography

Assmann, Aleida. "Canon and Archive." In *Cultural Memory Studies: An International and Interdisciplinary Handbook*, edited by Astrid Erill and Ansgar Nünning, 97–107. Berlin: Walter de Gruyter, 2010.

Barthes, Roland. *Image, Music, Text*. Translated by Stephen Heath. New York: Noonday Press, 1978.

Barthes, Roland. *Mythologies*. Translated by Annette Lavers. London: Grant and Cutler, 1994.

Berman, Sanford. *Prejudices and Antipathies: A Tract on the LC Subject Heads Concerning People*. Metuchen, NJ: Scarecrow Press, 1971.

Colomina, Beatriz. "Exhibition as Archive." In *The Archive as a Productive Space of Conflict*, edited by Markus Miessen and Yann Chateigné, 163–174. Berlin: Sternberg Press, 2016.

Derrida, Jacques. *Archive Fever: A Freudian Impression*. Chicago: University of Chicago Press, 1996.

Drake, Jarrett M. "RadTech Meets RadArch: Towards A New Principle for Archives and Archival Description." *Medium* (blog), May 5, 2018. https://medium.com/on-archivy/radtech-meets-radarch-towards-a-new-principle-for-archives-and-archival-description-568f133e4325#.86ba9wqwf.

Drucker, Johanna. "Le Petit Journal Des Refusées: A Graphical Reading." *Victorian Poetry* 48, no. 1 (2010): 137–169.

Drucker, Johanna. *The Visible Word: Experimental Typography and Modern Art, 1909–1923*. Chicago: University of Chicago Press, 1994.

Gardner, Eric. "Accessing Early Black Print." *Legacy: A Journal of American Women Writers* 33, no. 1 (2016): 25–30.

Giannachi, Gabriella. *Archive Everything: Mapping the Everyday*. Cambridge, MA: MIT Press, 2016.

Groth, Gary, Michael Catron, and Keeli McCarthy, eds. *The Complete Wimmen's Comix*. Facsimile edition. Seattle: Fantagraphics Books, 2016.

Howe, Susan. *Spontaneous Particulars: The Telepathy of Archives*. New York: New Directions, 2014.

Junghaus, Tímea. "Our Beloved Margins: The Imaginings of the Roma Transformative Subject and Art History Scholarship in Central Europe." In *Extending the Dialogue: Essays by Igor Zabel Award Laureates, Grant Recipients, and Jury Members, 2008–2014*, edited by Urška Jurman, Christiane Erharter, and Rawley Grau, 78–97. Ljubljana: Igor Zabel Association for Culture and Theory, 2016.

Kinross, Robin. *Modern Typography: An Essay in Critical History*. London: Hyphen Press, 1992.

Kleeman, Birte. "I Put This Moment Here." In *Wunderkammer*. edited by Birta Kleeman, 5–9. Brooklyn: Brooklyn Fine Arts, 2012.

Kronfeld, Chana. *On the Margins of Modernism: Decentering Literary Dynamics*. Berkeley: University of California Press, 1996.

Manoff, Marlene. "Theories of the Archive from Across the Disciplines." *Portal: Libraries and the Academy* 4, no. 1 (January 2004): 9–25.

Mullaney, Thomas S. *The Chinese Typewriter: A History*. Cambridge, MA: MIT Press, 2017.

Perneczky, Géza. *The Magazine Network: The Trends of Alternative Art in the Light of Their Periodicals, 1968–1988*. Köln: Soft Geometry, 1993.

Ram, Harsha. "Introducing Georgian Modernism." *Modernism/Modernity* 21, no. 1 (January 2014): 283–288.

Ridener, John. *From Polders to Postmodernism: A Concise History of Archival Theory*. Duluth, MN: Litwin Books, 2009.

Roberto, K. R. *Radical Cataloging: Essays at the Front*. Jefferson, NC: McFarland & Co., 2008.

Rooks, Noliwe. *Ladies' Pages: African American Women's Magazines and the Culture That Made Them*. New Brunswick, NJ: Rutgers University Press, 2004.

Scholes, Robert, and Clifford Wulfman. *Modernism in the Magazines: An Introduction*. New Haven, CT: Yale University Press, 2010.

Smith, Shawnta. *Black Lesbians in the 70's and before: An at Home Tour at the Lesbian Herstory Archives*. Brooklyn, NY: Lesbian Herstory Archives, 2010.

Spieker, Sven. *The Big Archive: Art from Bureaucracy*. Cambridge, MA: MIT Press, 2008.

Teige, Karel. "Umění dnes a zítra." In *Revoluční sborník Devětsil*, 187–202. Prague: Večernice V. Vortel, 1922.

Voss, Paul J., and Marta L. Werner. "Toward a Poetics of the Archive: Introduction." *Studies in the Literary Imagination; Atlanta, Ga* 32, no. 1 (Spring 1999): i–viii.

Williams, S. "Implications of Archival Labor." *Medium* (blog), April 11, 2016. https://medium.com/on-archivy/implications-of-archival-labor-b606d8d02014.

Part I

Image

1 Analog and digital Nigeria

Inheriting serial cultures in the work of Kelani Abass[1]

Jennifer Bajorek

Serial cultures of print and photography in West Africa

Histories of print culture in West Africa have often unsettled the paradigms of production and circulation made familiar by scholarship on print culture in other geographic and (geo)cultural sites and spaces. The practices of reading, writing, collective or collaborative literacy, audience participation, and DIY publication that have been vital to the emergence and critical elucidation of West African print cultures have yet to be adequately accounted for within the dominant theoretical frames, as work by scholars such as Karin Barber, Stephanie Newell, Onookome Okome, and Tsitsi Jaji has brilliantly demonstrated.[2] Important to underscore in a comparative (non-Africanist) context, this and other recent research on African print cultures, sited both in West Africa and in other parts of the continent, has worked to undo long-standing assumptions about print's status as a cultural praxis associated exclusively with urban elites in colonial and postcolonial space and has located print, instead, squarely within the arena of popular culture. Barber's research in particular has been instrumental in charting an explosion of printed matter that was produced and circulated not only by elites "but also by non-elites or obscure aspirants to elite status" – to such a degree that "local, small-scale print production became a part of social life" in urban West Africa.[3] Print's transcendence of communities defined by, and confined to, certain types of formal education (almost always in the colonial language) and associated literacies has reshaped our understanding of the relationships between print culture, European colonization, and local experiences of industrialization in colonial Africa. It has also cleared the way for new conceptualizations of both printed matter and popular culture and new theorizations of their role in processes and experiences of decolonization and in the formation of the postcolonial state.[4]

Much of this research has argued for a significant expansion of the very concept of printed matter in West Africa, redefining the "matter" of print culture to encompass domains extending far beyond those of literature, newspapers, or magazines and illuminating the complex interweaving of print technologies and printed objects with other popular cultural practices, media, and expressive forms.[5] Thus, West African print cultures have been shown by scholars such as Till Förster, Olubukola Gbadegesin, Achille Mbembe, John Picton, Kerstin Pinther, Ato Quayson, and Leslie Rabine to spill over into music (records), theater (photoplays), comics or cartoons, signboards, and sign-painting, and even fashion (wax-resist textiles and, more recently, streetwear, even T-shirts).[6] Much as the reframing of print culture as a species of popular culture has complicated received ideas about print's relationship to European colonization, the expansion of the concept of print to encompass a vastly richer field of cultural practices has gone hand in hand with a more general rethinking and reworking of earlier, and often overly binaristic understandings of popular culture. In a departure from previous constructions of popular culture as a homogeneous space of resistance to oppression, this new wave of scholarship has shed light on the plurality of social and political positions that popular cultures may both embody and engender.[7]

My own recent research, on mid-twentieth-century histories of photography in Francophone space (specifically, in Senegal and Benin), has led me to reflect on the intersection of photographic and print cultures in West Africa. In a move consistent with the above-mentioned expansions of the field of print culture, my own and others' research on printed images and, specifically, on photography has sought to loosen the stranglehold of the dominant (European and North American) paradigms and to point out their inability to account adequately for West African histories of the medium. The limitations of these paradigms are starkly apparent in those cases when they have relied on concepts of photography exclusively as a mechanical or technological process or have privileged its capacities for "mechanical reproducibility," understood narrowly within the logics of commodity production and of mass production dictated by industrial capital.[8] Processes of photographic manipulation, for example, have long played a starring role in West African photography histories, and practices such as drawing, writing, and painting manually on the surface of photographic prints with charcoal, graphite, crayon, ink, and gouache were widespread from the moment of photography's introduction in the region (a moment that coincided with that of the medium's invention). Another defining feature of these histories has been the production of multiples through serial rephotography – that is, by rephotographing a unique photographic print – rather than by striking multiple prints from a single negative through a negative-to-positive printing process. Unlike retouching of negatives (which West African photographers of course also did), and unlike the forms of digital manipulation that organize the dominant theoretical discourse on contemporary photography, manipulation in the form of manual alterations made on, or to, the surface of photographic prints and reproduction through serial rephotography both lead to the production of prints that are, strictly speaking, unique objects. These practices directly challenge the conceptual primacy of mechanical reproducibility as *the* concept grounding all theoretical work on photography, even as they call attention to the long-standing coexistence of manual and mechanical processes in what remain, precisely, serial expressions of the medium.

* * *

In what follows I explore a series of hypotheses about the interpenetration of manual and mechanical processes and practices in an attempt to break down the oppositions, or seeming oppositions, between manipulation and mechanical reproduction that I have just sketched. After all, manual and mechanical processes have always coexisted in both printing and in photography, yet I believe that theorization of their coexistence has often been clouded by a preoccupation with mechanical reproducibility. As the inclusion of a particular photograph of Fatou Ba in the illustrated magazine *Bingo* in 1953 makes clear, the magazine's editors had no qualms about printing photographs that had been manipulated through processes of manual retouching and overpainting that had been developed in West African studios over a century of local portrait traditions. Images such as the overpainted photograph of Ba (the leader of a local socialist women's committee) brought recognizably West African aesthetics and visual codes onto *Bingo*'s pages, even as the magazine promoted an often explicitly consumerist vision of technological modernity.[9] The magazine's readers clearly also saw no contradiction in this marriage of seemingly more artisanal or craft-based aesthetics with those native to a mass circulation serial publication: they also routinely submitted portraits of themselves that had been hand-colorized, inked, or overpainted for publication in the magazine.[10] These reader-submitted portraits appeared alongside commissioned photo essays, contemporary photojournalism, and reportage on topics ranging from the war in Indochina or new infrastructure projects (dams, railroads, bridges, modernization of ports) to the "African independences" in 1960, or in features that were illustrated, in the late 1950s, by composite manipulated photographs of the Tidjane saints.[11] With these examples, I want to underscore that manual and mechanical processes of production and reproduction have happily coexisted as an integral part of West African histories of both print and photography. What if we acknowledged this coexistence more fully than it has been in the scholarship to date? Would this not alter the dominant paradigms for thinking about both print and photography – in West Africa and in the rest of the world?

This chapter pursues these questions through an interpretation of work by contemporary artist Kelani Abass. Abass's work often explicitly activates and calls our attention to the interplay of manual and mechanical processes with regard to both printing and photography, combining or recombining, with painting, the aesthetic and material products (texts, images) of different technologies of printing, copying, and writing. The artist's most recent work adds digital printing and imaging technologies to the mix, translating the experiential dimensions of texts, images, and objects associated with an array of different printing and photographic technologies across distinctly different media and periods of time. Abass's work offers, I will suggest, a distinctly West African vantage point both on the history of print and of photographic cultures and on their present-day global transformations.[12]

Kelani Abass: recent work

Starting in 2011, in an exhibition of his work at Omenka Gallery in Lagos, Nigeria, titled *Man and Machine*, Abass began to show paintings into which he had integrated images of mechanical and technological objects as well as fragments of printed text: references to the time that he spent, before going to art school, as a machine operator in his family's commercial printing press in Abeokuta, Nigeria.[13] In a move that was not exactly a turn away from painting – a medium in which the artist continues to work, along with photography, assemblage, collage, and a range of printing and stamping techniques – Abass henceforth experimented more actively with directly incorporating other media into his paintings. In 2013, in another solo exhibition titled *Asiko* curated by Bisi Silvia at the Centre for Contemporary Art–Lagos (CCA–Lagos), Abass included photographs and printed texts, as well as typographic and iconographic references to techniques, processes, and printed objects – such as illustrated calendars detailing the dates of accession of local Yoruba kings – connected with his family's printing business in Abeokuta (Figures 1.2–1.4).[14] In the largest paintings that were exhibited in *Asiko*, photographs from Abass's family album were enlarged and laminated to the canvas in mosaic arrangements approximating a grid.[15] In *Asiko* (2012), *Asiko 2* (2013), *Ale* (2013), and *Irole* (2013) – all part of the *Family Album* series (Figure 1.2) – the artist combines reproductions of vintage black-and-white prints, corrugated cardboard, painting (acrylic on canvas), and visual references to digital imaging technologies.[16]

Abass's latest work, exhibited in another solo show at the CCA–Lagos, also curated by Silva in late 2016 and early 2017, titled *If I could save time . . . I'd just roll ink on the paper*, manifests an even more acute preoccupation with ordering, reordering, and arranging textual, visual, typographic, and iconographic elements connected with the printing press and related techniques, processes, and objects (Figures 1.5 and 1.6).[17] Here, typographic, iconographic, and other aesthetic and material residues of printing machines, printing-related equipment, and printed matter have been both embedded into the two-dimensional canvas and installed as part of three-dimensional wall-mounted sculptures. In a work from the series from *If I could save time . . .*, titled *Casing History*, Abass's wall-mounted sculptures make use of mosaics and semiregular gridlike arrangements similar to those that we saw on canvas in paintings such as *Asiko 2* (Figures 1.2 and 1.5). In the *Casing History* series, however, the rectilinear forms are visually resonant with those of letterpress typecases, and actual typecases into which photographs and textual objects – including a manuscript journal – have been fitted into compartments of varying size have been mounted to the wall. In another work in *If I could save time . . .*, titled (*Stamping History*) *Making Time 1* (2015–2016) (Figure 1.6), a group portrait from Abass's family album is reproduced, in 16 contiguous panels at monumental dimensions (approximately 8 × 12 feet), through the stamping of cardinal numbers using a manual hand-numbering machine.[18] I will return to this work at greater length below.

Among the individual works exhibited in *Asiko* were two paintings (acrylic and oil on canvas) depicting letterpress printing presses and a painting of a mimeograph machine dating from the 1950s.[19] These "portraits" of printing presses and other copying and writing machines are framed by vignette-style borders, underscoring their status as portraits and invoking in painting the material and aesthetic qualities of printed objects (Figure 1.1: *Family Portrait 1*). These portraits find their counterparts in a series titled *Family Album*, which consists of large-scale canvas diptychs and triptychs featuring reproductions of photographs from Abass's own family albums (Figure 1.2). The artist has scanned vintage black-and-white prints dating from the 1960s and 1970s and enlarged and reprinted them at generous dimensions using digital technologies. The new photographs have been over-painted and, in many cases, collaged with other photographs. The crosshairs of digital viewfinders, numbers indicating timecodes, color-matching charts, and lines evoking various editing tools have, in some works, been superimposed on the photographs, repeating but also adding a new layer of rectilinear pattern and superimposing the hues, and feel, of looking at a monitor or screen on the experience of looking at the canvas. In many of the photographs that have been included in the works in the *Asiko* exhibition, it is possible to discern repeat instances of the same man and woman, whose outlines we also recognize sketched in acrylic elsewhere on the canvas: Abass's parents (see, again, Figure 1.2: *Asiko 2* diptych).

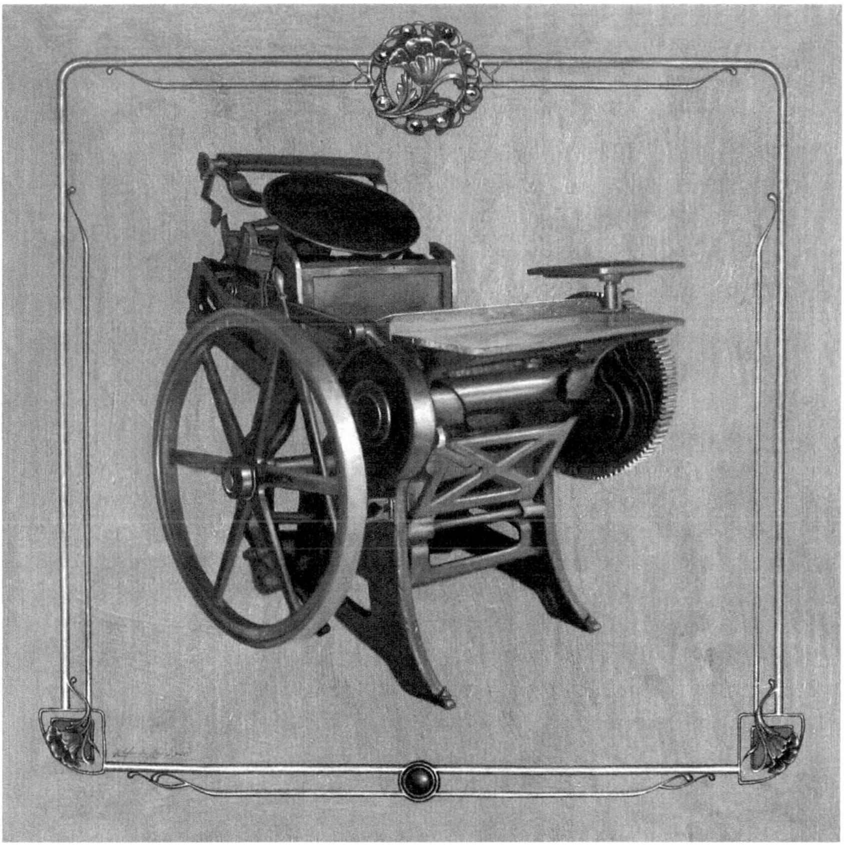

Figure 1.1 Kelani Abass, *Family Portrait 2* (2013), acrylics and oil on canvas, 122 × 122 × 6 cm.

Source: Reproduction courtesy of the artist.

The works comprising the *Asiko* and *If I could save time . . .* exhibitions confront the viewer with insistent questions about the role of specific technologies in duration and memory, archiving, and inheritance. The paintings and photographs in the *Family Portrait* and *Family Album* series frame these questions across a dizzying array of materialities associated with diverse moments in the history of print culture, incorporating images of – and graphic, textual, visual, and aesthetic references to – presses as well as typecases, Gothic lettering as well as time code, black-and-white prints as well as digital photographs. The artist's decision to position the image of a printing press and other copying and writing machines as members of a family, and, it is implied, as members of *his* family – and thus as ancestors from which he is descended in a direct line – within the *Family Portrait* series, unmistakably positions both the artist and his work as an inheritor of these machines and of the objects that they produced. The artist's decision to claim this inheritance is made in tandem with a second decision, to make explicit the links between printed objects or textual cultures and local political and social histories, as represented by the dates of accession of local monarchs represented on the printed calendars or by the manuscript journal in which local knowledge and family history have been captured and handed down from generation to generation. The work's persistent claim to a printed inheritance that is at once personal and collective or communal troubles long-standing and problematic assumptions about European or North American monopolies on industrial, capitalist, and colonial as well as postcolonial modernities.

Visual inheritance

Abass's parents figure conspicuously in the canvases in the *Family Album* series. In *Asiko 2*, photographs from Abass's family album have been reproduced and arranged into two different groups. The diptych format enhances our awareness of existing patterns, both symmetries and asymmetries, in the arrangement of the photographs. At the same time, this format gestures towards the existence of a much larger archive of images than those that are actually visible to us. This feeling, that there are still more images beyond those we see in the work, is amplified by the addition of vivid red horizontal and vertical lines, suggestive of the crosshairs of a digital viewfinder, or perhaps of the *x* and *y* coordinates of a digital cropping tool or cursor in Photoshop. The obscuring of some photographs beneath a thick layer of acrylic paint serves, in some cases, to foreground the faces and individual identities of the photographic subjects. In other cases, this overpainting serves to render the faces of the subjects hardly visible. In the first instance, the paint seems to preserve, and call our attention to, these individuals' identities in part by rendering the details of the framing event, occasion, or gathering more abstract. In the second instance, the obscuring of the photographic subjects' faces seems to suggest that their identities are not the work's principal concern. In this case, the technique of overpainting is used to suggest that, in fact, all we really need to know is that there was a photograph. This technique of overpainting, here used by Abass, is consistent with West African photography histories that have

Figure 1.2 Kelani Abass, *Asiko 2 (Family Album Series)* (2013), diptych, corrugated cardboard, laminated print, and acrylics on canvas, 91 × 122 × 6 cm each.

Source: Reproduction courtesy of the artist.

veered sharply away from a concern with photographic indexicality (construed, almost overwhelmingly in Western/Northern contexts, in a positivist sense, producing an overemphasis on fixity and reality effects) to underscore the relentless mutability of the photographic object. As Erin Haney asks, eloquently, with regard to the significance of this mutability to West African traditions: "Are these images [i.e., photographs] so treasured because of their connection to a lived presence or because of the possibilities they offer as mutable objects?"[20]

The reproduction and superimposition of some of the figures, sketched in acrylic, at significantly larger dimensions in contrasting colors, urge us to reflect, on yet another level, on the significance of particular individuals' identities. It also invites us to consider the status of their portraits (whether painted or photographic) as material objects. The painted figures may seem at first to draw our gaze away from, but then to return it to, the photographs. This second rendering, or doubling, of some of the figures in acrylic invites us to speculate about the relationships of these individuals to the other people depicted in the images. That some of the painted figures appear to turn their backs only heightens our awareness of their uncertain status as photographic or quasi-photographic subjects, and raises still more questions about their relationship to the viewer: Are they turning their backs on the camera or to us? Are they turning their backs on the camera *and, therefore*, to us? Upon closer inspection, it proves more difficult than we may at first have imagined to match up the painted figures directly with those depicted in the photographs: in several cases, the painted figures are clearly the same people we discern in the photographic portraits, yet the painted figures appear not to be exact reproductions of the photographed subjects. Or, if they are reproductions, some line or gesture is always highlighted or exaggerated, such that it appears to take on greater significance. In view of this subtle yet insistent highlighting of particular lines or gestures, it is almost as if we are seeing the image of the individual in memory rather than in a photograph. As if to foreground the role of memory in the interpretation of images, the viewer is invited to recognize patterns: if not precise repetitions of lines and forms, then the recurrence of certain visual motifs, such as a particular handbag or a way of holding it, a particular pair of earrings, a particular way of tying a headscarf. The seemingly casual arrangement of the painted figures, in contrast to the "poses" of those depicted in the photographs,

Figure 1.3 Kelani Abass, *Memoirs (Calendar Series)* (2012), corrugated cardboard, laminated print, charcoal and acrylics on canvas, 122 × 122 × 6 cm.

Source: Reproduction courtesy of the artist.

Figure 1.4 Kelani Abass, *Oba Gbadebo I (Calendar Series)* (2013), corrugated cardboard, laminated print, charcoal and acrylics on canvas, 105 × 122 × 6 cm.

Source: Reproduction courtesy of the artist.

creates an impression of spontaneity. Perhaps they have gone back to other conversations – and have been only momentarily interrupted by the photographer.

* * *

The work in the *Asiko* exhibition included, together with photographs drawn from Abass's family album, photographs of other people – royalty, dignitaries, or other local notables – that were printed in a series of illustrated calendars, using letterpress technology, by his father (Figures 1.3 and 1.4).[21] The inclusion of these particular images is important because it brings the larger community in and around Abeokuta into Abass's image repertoire, and because it joins portraits of the artist's immediate family (again, we see portraits of both parents and especially of his father) to those of Yoruba monarchs (in the type of calendar that Abass identifies as "Bomode Oku": Figure 1.4) and other prominent members of the community into a self-consciously serial form: the calendar.[22] This inclusion also brings the labor of prior generations of local printers, and multiple printing and photographic processes, into the mix. In an interview that the artist conducted with scholar and curator Jude Anogwih, published in the *Asiko* catalogue, Abass explains that the photographs reproduced in the "Bomode Oku" calendars and reproduced by the artist, again, in the *Calendar* series were "derived from an old metal plate used in reproducing photographs in letterpress printing."[23] Abass adds that these plates were silver-coated and treated with acid to make them sensitive to light, and that these photographs were thus reproduced using a technique that is reminiscent of early photographic technologies such as the daguerreotype, and therefore considerably different from those used to make many of the photographs that are also incorporated into the work (most of which would have been taken with medium-format cameras using tripods and, later, smaller format cameras such as the twin lens reflex or the 35mm).

Through this mixing of different printing and photographic technologies, the works in the *Calendar* series revisit the calendar as a historical object. As Abass explains, these calendars of the "Bomode Oku" variety, in which photographic portraits of Yoruba monarchs are printed alongside the dates of accession to the throne of their forebears, demonstrate both the importance of the monarchy to local culture in Abeokuta and the deep integration of its rhythms of succession into daily life. More specifically, Abass notes that the calendars demonstrate a particular concern, in Abeokuta, with documenting transitions of power. Thus, in one of the works in the *Calendar* series, *Olori Nlado* (2012) (not reproduced here), a portrait of a woman (the photographer's grandmother) is bounded, on the right, by a calendar in the familiar grid format for the month of January 1975 and bounded, on the left, by an exhaustive list of the dates of accession for each oba for the date range 1854–1972. Another work in the *Calendar* series, *Oba Gbadebo I* (2013) (Figure 1.4), frames its central photograph with a list of "firsts" in Abeokuta's history, such as (to the left), dates for the "emancipation of slaves" (1834), for the first persecution of a Christian in Abeokuta (1848), and for the first commitment of the Yoruba language to writing (also 1848). The calendar is for the month of May 1904 and could not have been printed by Abass's father, yet it links the artist to a longer tradition of local artisans, printers, and keepers of collective memory in the wider community.[24]

In his interview with Anogwih, Abass emphasizes the extent to which the layout and design of several works in the calendar series emulate the layout of a page for offset printing.[25] That is to say, more specifically, Abass's process combines, technically and visually, elements from the calendars with new elements and rearranges them on the "page" (canvas); he paints over them with acrylic; he scans and enlarges the photographs to create digital versions; decorative borders are added (Figure 1.3). In other works from the *Asiko* series (see, again, *Asiko 2*: Figure 1.2), Abass extends the rigorous procedures of dating that he also explores in the calendar series by "providing" the time code and digital date stamp that would have corresponded, could the materiality of the technologies be confronted more directly, to the vintage photograph.

A similar preoccupation with ordering, reordering, arranging, and rearranging textual, visual, typographic, iconographic, aesthetic, and material elements of

Figure 1.5 Kelani Abass, *Casing History 1* (2016), letterpress typecase and digital prints, 36 × 82 cm.

Source: Reproduction courtesy of the artist.

different printing technologies can be seen in the works in another, more recent series by Abass, titled *Casing History*. The works in *Casing History* (Figure 1.5) were produced in 2016 and exhibited, again, at the CCA–Lagos in the 2016–2017 exhibition, *If I could save time . . .*, mentioned above.[26] In the works in *Casing History*, instead of arranging text and image as if to prepare a page for offset printing, and in ways reminiscent of even earlier printing technologies, Abass has (as I mentioned in passing already) inserted the photographs into a letterpress typecase. The typecase has been turned on its side and operates like a frame, or like a honeycomb of smaller frames, arranged, again, in a kind of grid formation. Artist and curator Temitayo Ogunbiyi, in an insightful essay published in the exhibition catalogue, has noted the range of tones represented in the prints that Abass has reproduced through digital techniques: she refers to "uncalibrated palettes of grayscales, sepias, and greens."[27] It is as if the painter no longer needed his paint, with the hues present in the photographs now providing his only palette.

Vitally, *Casing History 1* mixes manuscript objects alongside photographs in the typecase. In this particular work we see a postcard as well as pages from a handwritten journal recording Abass's family history, which, he explains, has been passed down through generations for over 70 years. Also in her catalogue text, Ogunbiyi describes the systematic logic governing the arrangement of the images in all of the works in the *Casing History* series: the compartments used for uppercase letters are the smaller ones, and these are the compartments selected by Abass to house the single portraits of individual sitters; the compartments used for lowercase letters are the larger ones, and these are the compartments selected by Abass to house group portraits, which have in several cases been cut into sections so that they can be fit into consecutive compartments.

Not surprisingly, Abass's reasons for this distribution are deeply connected with the artist's ideas about memory, archiving, and collective experience. He explains these reasons at some length in another interview, published in the *If I could save time . . .* catalogue, with Odun Orimolade. In printing, Abass suggests, the smaller letters, the lowercase letters, are by far the most important: they look small, but a word or a sentence depends upon many small letters grouped together in order to make its meaning; the capital letter is, despite its larger size, in fact comparatively insignificant in this process. More specifically, Abass states: "I realize that the part that capital letters play in making a sentence is less significant to that of small letters."[28] Just as, Abass offers, the individual is insignificant in history: "In essence, the impact we make individually cannot be compared with what we can achieve collectively."[29] The work is, in part, a meditation on multiple technologies and supports of memory, and it is organized by multiple archival practices and logics: those proper to photography (and, within photography, appropriate to individual and group portraits) are joined to those proper to letterpress printing, and both are joined, in turn, to those proper to the manuscript or handwritten text. At the same time, however, the work is clearly driven by an impulse to visualize or materialize the radically *collective* nature of any archive, act of archiving, or repository of memory. Here, the arrangement of lowercase and uppercase letters in the letterpress typecase and the logic of alphabetic language and specifically of the Roman alphabet, as it circuits through the printing press, finds its mirror image in the arrangement of photographs in every (any) family album. Both the family portrait or the family album (featuring images of the individual and of the collective) and the typecase (featuring both the capital as well as the lowercase letter) are here joined in an allegory of the radically collective nature of history, and, simultaneously, of the radically social dimensions of memory or of archival practice.

Serial inheritance

In a recent e-mail exchange that I had with the artist, he underscored one further dimension of the work in the *Casing History* series.[30] When I asked him about the gridlines that run, in fact, not only through *Casing History*, but arguably through all of this most recent work (that from *Asiko* as well as *If I could save time . . .*), he folded the gridlines directly back into a meditation on acts and practices of ordering, reordering, arranging, selecting, and archiving, suggesting that the people in the images experienced some of these processes already, and were subjected to them already, in the moment that they were photographed: "What really informed [the] grid in those pieces is the idea of photographic technology and how it has affected the way we see ourselves, how we present ourselves, how the availability of cameras and camera-enabled devices accessible to us have made us photocentric." He continues in his e-mail: "There is a lot of digital photocentricity in our world today," adding that, "Where the camera exists, almost everyone can afford one, and nobody can imagine life without it." But, Abass suggests, by contrast, this was not the case for the people depicted in the images, including his parents, whose relation to photography, although they undoubtedly had one, was clearly different from our own. "The grid, he says, represents life inside the camera focus for those people," who lived and who were photographed at a time when there was not yet widespread access to cameras. When viewed in this light, Abass's *Casing History* works allow us to "remember," to imagine, to see and to feel the experiences and practices of publishing, of remembering (or commemorating), and of archiving that date from a time when we were not yet so photocentric. Counterintuitively, perhaps, it does so precisely through exploring photography's ongoing evolution and its (ever-evolving) differences from itself. "Photographs," he says, "have the effect of bringing distant memory to [the] surface."

* * *

The work titled *Making Time 1* (*Stamping History*) (2016), the final work by Abass that I will discuss, makes a similarly provocative comment on our own photocentricity, and how it might allow us to reconsider both our technological difference from, and similarity to, the past (Figure 1.6). Using what Ogunbiyi describes, in her catalogue essay, as a photorealist technique, a group portrait from Abass's family album is reproduced in 16 contiguous panels, through the stamping of cardinal numbers (advancing, in principle although not in practice, to infinity) using a manual hand-numbering machine. In all of the texts published in the catalogue, Abass and his interlocutors have taken care to emphasize the embodied

Figure 1.6 Kelani Abass, *Making Time 1* (*Stamping History Series*) (2016), manual hand-numbering machine on paper, 254 × 363 cm.

Source: Reproduction courtesy of the artist.

Figure 1.7 Kelani Abass, *Making Time 1* (*Stamping History Series*) (2016), detail.

experience – and indeed the labor – invoked by this machine by the artist, and he recalls the many long hours that he spent as a child stamping invoices and receipts in the afternoons in his family's printing press, using the manual hand-numbering machine.

Abass describes this labor as tiring – it was physically difficult work – and also as tedious. Manual stamping was, and is, a continuous and repetitive activity. It is from this memory of stamping, and of numbering pieces of paper, that the title of the exhibition appears to have been derived: *If I could save time . . . I would just roll ink on the paper* – rolling would be, presumably, easier and less tedious than stamping. Abass says to Orimolade in their interview,

> To be emphatic about a mark, and to keep score at the same time through the numbering process, it testifies to contact between one [a person] and the machine. In the process of stamping, there were periods I felt like speeding up time, or look[ing] for something that could just roll over it and make impressions on the paper for me, because of the enormity of the work.[31]

I should add that, in this description, the artist seems (to me) to be talking at one and the same time about the experience of stamping as a child in the family printing business in the afternoons after school and the experience of making the work today, in 2016, which ironically employs the same mode of production that he found tedious as a child.

In this iteration of this manipulated photograph – one produced through both manual and mechanical processes – we can almost see the past and the future meet in the present with which we are also confronted by the work. This meeting is no doubt (yet this fact is brought into particularly sharp relief here) constitutive of all serial forms. Past and future are here merged both in the technologies of printing and other processes that the artist uses in the creation of the work and in the embodied relationship that he has to these technologies and processes, as much as in the object that results (Figure 1.7); for instance, in the palimpsest of a photograph, a line drawing, and numbers stamped in ink, or in the embodied memory of stamping and the forward moving process of numbering in a potentially infinite series.[32] The artist describes the moment of stamping as, again, a moment of contact "between oneself and the machine," a moment that continues to shape and inflect his relationship to his family, and to the memory of his father (the printer) in particular.

I have referred to this work as a manipulated photograph, but, in his description of the process of making this work in our e-mail exchange, Abass refers several times to this particular work, *Making Time 1*, as a *drawing*. I find this word helpful, as it clearly refers to specific aspects of the artist's process in the work: it was made by the application of ink, by hand, to paper (a definition of drawing, although it is also a definition of stamping); as he underscores in his description, he also created a line drawing from the photograph, over which he then stamped. The layering of these processes is part of what *Making Time 1* presents to us as, and in the context of, a meditation on seriality: a black-and-white photograph (a vintage or so-called analog print) is scanned and enlarged using digital technologies; this digital image then becomes the basis of a line drawing, which then becomes the basis of the stamping or printing process, and thus another printed object. As such, the word "drawing" serves to underscore the coexistence of manual and mechanical processes that were already constitutive of the hand-numbering machine, and perhaps already present in the name of the thing: "hand-numbering machine." Abass describes the process of the work's production, in his e-mail, thus:

> The process starts with a line drawing after which the numbering machine is introduced to strengthen the lines giving it a three dimensional look, this occurs by the stamping of numbers in ascending order, which is done continuously until the drawing is finally complete. An interesting feature of the numbering machine is that it doesn't duplicate numbers, it keeps printing numbers as you make an impression with it as in the case of a receipt, unless set to duplicate.

With this comment, Abass thus brings out the many parallels between this technology, leading in principle to an infinite series, photography, and all serial forms. Add to which, it is interesting to think that we could ever use this or any other technology so deeply bound up with an unbound seriality to produce (or reproduce) something that feels "complete."

This brings us back to Haney's question about the inherent manipulability of photography, which is precisely what opens it to memory: "Are these images so treasured because of their connection to a lived presence or because of the possibilities they offer as mutable objects?"[33] Haney's question helps us to frame more clearly the opening to memory of both printed image and printed text that I have sought to describe in Abass's work. As we have seen, his work effects this opening through a whole host of techniques of reproduction and manipulation that suggest new vocabularies for conceptualizing printed matter and serial cultures in accordance with Nigerian experiences and histories. Many of these techniques combine both manual and mechanical processes of manipulation. As such, they challenge the conceptual primacy of a purely mechanical reproducibility – thought, in the West and North, to have defined both photography and printing press. At stake is not simply this challenge to a master trope that, as such, cannot exhaustively define the history of technology (as if any technology could have a single history) but the collective and social dimensions of the experiences and histories that are traced in, and transmitted by, these printed objects.

Notes

1 This chapter started life as a paper presented at the Arts Council of the African Studies Association (ACASA) Triennial Symposium on African Art, at the University of Ghana, Legon, August 8–11, 2017. I am grateful to Érika Nimis and Marian Nur Goni, organizers of the double panel on "Handling/Manipulating Photographs in Africa," for including me and to the other panelists and audience members for their questions and comments on the paper. I am also grateful to Erin Haney, Nora Kennedy, Peter Mustardo, Bisi Silva, and Abraham Oghobase, for the conversations that we had with the artist about his work on a trip to Lagos in April 2014. Above all, I am grateful to the artist, Kelani Abass, for sharing his work during that same visit, for granting me permission to reproduce photographs of his work, and for responding so generously to my interview questions, by e-mail and WhatsApp, after the fact.

2 Karin Barber has produced a particularly compelling body of research on what she calls "tin-trunk literacy" and "DIY modes" of editing and publishing in West Africa, and she has argued forcefully for a definition of West African print cultures that were not confined to highly educated elites. See, in particular, her introduction to *Africa's Hidden Histories: Everyday Literacy and Making the Self* (Bloomington, IN: Indiana University Press, 2006). See also Stephanie Newell and Onookome Okome's excellent edited volume, *Popular Culture in Africa: The Episteme of the Everyday* (London and New York: Routledge, 2013); and, more recently, Tsitsi Jaji, *Africa in Stereo: Modernism, Music, and Pan-African Solidarity* (Oxford: Oxford University Press, 2014), which treats *Bingo*, the West African illustrated magazine, in a comparative reading with *Drum* (an illustrated magazine published in South Africa), thus following the complex circuits traced by popular music, print cultures, and photography in multiple contexts in postcolonial Africa.

3 Barber, "Introduction: Hidden Innovators in Africa," in *Africa's Hidden Histories*, 1. Barber adds that, in coastal West Africa, "pamphlets and booklets were run off on artisan presses by entrepreneurs who were sometimes also the editors and even the authors of the texts they printed" (1).

4 Jaji's research is particularly lucid in its exploration of the links between commercialization, commodification, and the inculcation of a range of media literacies in postcolonial Africa, and she writes, specifically, about efforts to train readers, through magazine literacy, in "a new set of interarticulated desires for new products and states of mind" (Jaji, *Africa in Stereo*, 115).

5 Prior to their redefinition in the expanded field, local print cultures were generally understood as the local production of printed documents intended for consumption by local audiences. According to this definition, printed documents or objects produced in Europe and exported to West Africa for a European and/or African audience would not be enough to constitute a local print culture, although such documents might arguably help to create the conditions for one. Adding yet another wrinkle to the definition of print culture in the region, photography itself can and should be seen as part of local print cultures in West Africa, even if, in many parts of the region, local photographic traditions actually preceded the local production of printed texts. On the different rhythms of print culture's emergence in Francophone as compared with Anglophone West Africa, see Thierry Perret, *Le temps des journalistes. L'invention de la presse en Afrique francophone* (Paris: Karthala, 2005); Gil-François Euvrard, "La presse en Afrique Occidentale Française des origines aux indépendances et conservée à la Bibliothèque Nationale" (Mémoire, École Nationale des Bibliothèques, France, 1982), 15–16; Rosalyn Ainslie, *The Press in Africa: Communications Past and Present* (New York: Walker and Company, 1966), 130.

6 In addition to the work by Barber, Newell, Okome, and Jaji cited above, see, on signs and sign-painting, Till Förster, "Visual Presence and Competition in Urban Africa," *Critical Interventions, Journal of African Art History and Visual Culture* 1, no. 2 (2008): 59–79 (Cameroon and Côte d'Ivoire) and Ato Quayson, "Signs of the Times: Discourse Ecologies and Street Life on Oxford St., Accra," *City & Society* 22 (2010): 72–96 (Ghana); on textiles, John Picton, "Colonial Pretense and African Resistance, or Subversion Subverted: Commemorative Textiles in Sub-Saharan Africa," in *The Short Century: Independence and Liberation Movements in Africa, 1945–1994*, ed. Okwui Enwezor (München: Prestel, 2001) (Ghana), and Kerstin Pinther, "Textiles and Photography in West Africa," *Critical Interventions* 1, no. 1 (2007): 106–118 (Ghana); on streetwear and T-shirts, Leslie W. Rabine, "Remapping the Worldview of Fashion," *International Journal of Fashion Studies* 3, no. 2 (2016): 185–208 (Senegal); and, on photoplays, Olubukola Gbadegesin, "Photography and Performance in the Yoruba Photoplay Series," Unpublished paper presented at the Arts Council of the African Studies Association (ACASA) Triennial Symposium on African Art, August 9, 2017 (Nigeria). On political cartoons in postcolonial Cameroon, see Achille Mbembe's treatment of this topic in his well-known book, *On the Postcolony*, specifically Chapter 3.

7 Sidney Kasfir has posited that there are essentially two competing schools in theories of popular culture in Africa: one that defines popular culture as a space of contestation (if not necessarily of resistance or of liberation) and one that argues for the "emergent, unbounded qualities of popular culture" and arrives at a more fluid understanding of its capacities. See Kasfir's review of Johannes Fabian, "Review of Johannes Fabian, *Moments of Freedom: Anthropology and Popular Culture*." *American Anthropologist* 102, no. 4 (2000): 937–938.

8 The idea that the mechanical reproducibility of photographic images is essential to the medium of photography is, today, often associated with the German-Jewish cultural critic and philosopher Walter Benjamin, who formulated a powerful argument to this effect in his 1935 essay, "The Work of Art in the Age of Its Technological Reproducibility," in *Selected Writings, Volume 3 (1935–1938)*, trans. Edmund Jephcott and Harry Zohn, ed. Michael W. Jennings (Cambridge, MA: Harvard University Press, 2006), 101–133. However, versions of this idea have been cherished in theories of photography from the inception of the medium, and one of photography's inventors, William

Henry Fox Talbot, described photography as "the pencil of nature." See Fox Talbot, "A Brief Historical Sketch of the Invention of the Art," in *Classic Essays on Photography*, ed. Alan Trachtenberg (New Haven, CT: Leete's Island Books, 1980), 29.

9 This vision was also, I demonstrate elsewhere, a colonial one, and ultimately a decolonial one. See Chapter 3 of my book, Jennifer Bajorek, *Unfixed: Photography and Decolonial Imagination in West Africa* (Durham, NC: Duke University Press, 2019), which examines, in particular, the magazine's policy of soliciting and publishing photographs from its readers.

10 See, again, Chapter 3 of my book.

11 One such composite image, comprised of a cameo-style portrait of the famous Tidjane saint, El Hadj Malick Sy, surrounded by portraits of his descendants, was almost certainly intended as a devotional or quasi-devotional image when it was produced, but was published in *Bingo* in a didactic context. The Tidjanes are one of the three major Muslim *confrèries*, or brotherhoods, in Senegal.

12 In choosing to focus on a Nigerian artist, I am straying from the geographic and cultural focus of my previous research on photography in Francophone space. As mentioned above (in note 5), the emergence of local print cultures followed a different rhythm in Francophone as compared with Anglophone West Africa. There are nonetheless also many similarities and points of contact – aesthetic and otherwise – within print and photographic cultures across the region. These similarities are not an explicit theme of this essay, which focuses on Abass's work.

13 Abass trained at the Yaba College of Technology, where he did his first degree in General Art and a subsequent degree in painting, receiving a Higher National Diploma in painting, with distinction, in 2007. *Man and Machine*, the 2011 exhibition at Omenka, was his second solo exhibition. This description of *Man and Machine* is a paraphrase of Bisi Silva's description, published in Kelani Abass, *Asiko: Evoking Personal Narratives and Collective History* (Lagos: CCA–Lagos, 2013) (the catalogue of the *Asiko* exhibition), 7.

14 I did not visit the exhibition in person but was able to look at documentation of the exhibition and related work with the artist during a trip to Lagos in April 2014, and also to see a handful of the paintings that were then displayed in the library of the CCA–Lagos, as well as to consult the exhibition catalogue: Abass, *Asiko*.

15 The largest canvases were approximately 60 × 72 inches (dimensions are given in centimeters in the exhibition catalog) for each canvas in the diptych, meaning that the combined dimensions were as large as 10 × 12 feet.

16 Kunle Filani notes that Abass does not just reproduce these black-and-white photographs but also restores them before rephotographing, enlarging, and laminating them. Filani, "Memoirs and Memories: Kelani Abass in the Crucible of Conceptual Creativity," in Abass, *Asiko*, 16.

17 Exhibition catalogue published as Kelani Abass, *If I Could Save Time . . . I'd Just Roll Ink on the Paper: November 19, 2016–January 21, 2017* (Lagos: CCA–Lagos, 2016).

18 Ibid., 6–11.

19 Also included in the *Asiko* catalogue was an installation shot of a typewriter used for an interactive performance involving the artist's mother in the exhibition space. The artist invited his mother to do an interactive performance: using the same typewriter that she used between 1960 and 1970, she typed what she experienced with the artist's father, and how they both started the business. (His mother worked primarily as a typist, producing typed documents from manuscripts brought to her by clients, while his father operated the presses.) The audience then participated in typing their personal experience after she was done. Personal communication with the artist, May 7, 2018.

20 Erin Haney, "Film, Charcoal, Time: Contemporaneities in Gold Coast Photographs," *History of Photography* 34, no. 2 (May 2010): 120.

21 "In Conversation: Kelani Abass and Jude Anogwih," in Abass, *Asiko*, 26.

22 I will return to these "Bomode Oku" calendars later in this essay.

23 Ibid. Abass is here describing the process by which these photographs were originally reproduced by printers, including his father, producing the "Bomode Oku" calendars.
24 The artist explained that the "Bomode Oku" calendar was a tradition common amongst families in Abeokuta, and that they were commissioned by printers as an end-of-year souvenir gifted to clients, friends, and relatives. He used a "Bomode Oku" calendar drawn from a local archive as a template for the works in the *Calendar* series, selecting text from the history of the monarchy of Egba Land and adapting it for *Oba Gbadebo I* (Figure 1.4). [Personal communication with the artist, May 7, 2018.]
25 "In Conversation: Kelani Abass and Jude Anogwih," in Abass, *Asiko*, 27.
26 There are 20 works in the series titled *Casing History* that were reproduced in the *If I Could Save Time . . .* catalogue.
27 Temitayo Ogunbiyi, "In Time for History," in Abass, *If I Could Save Time . . .*, 24.
28 "Remembering to Remember: Objects of History and Memory (Kelani Abass in discussion with Odun Orimolade)," Abass, *If I Could Save Time . . .*, 40.
29 Ibid.
30 Personal communication with the artist, August 9, 2017.
31 "Remembering to Remember: Objects of History and Memory (Kelani Abass in discussion with Odun Orimolade)," in Abass, *If I Could Save Time . . .*, 41.
32 As Orimolade astutely notes, the capacity of the stamping tool to generate both images *and* narratives is activated by the work. He states: "I find it interesting that you are generating new imagery from the stamping tool, a printing tool that becomes the subject of your narrative as opposed to using imagery from the archive" (Ibid., 40).
33 Haney, "Film, Charcoal, Time," 120.

Bibliography

Abass, Kelani. *Asiko: Evoking Personal Narratives and Collective History*. Lagos: CCA-Lagos, 2013.

Abass, Kelani. *If I Could Save Time . . . I'd Just Roll Ink on the Paper: November 19, 2016-January 21, 2017*. Lagos: CCA-Lagos, 2016.

Ainslie, Rosalynde. *The Press in Africa: Communications Past and Present*. New York: Walker and Company, 1966.

Bajorek, Jennifer. *Unfixed: Photography and Decolonial Imagination in West Africa*. Durham, NC: Duke University Press, 2019.

Barber, Karin, ed. *Africa's Hidden Histories: Everyday Literacy and Making the Self*. Bloomington, IN: Indiana University Press, 2006.

Benjamin, Walter. "The Work of Art in the Age of Its Technological Reproducibility." In *Selected Writings, Volume 3 (1935–1938)*, translated by Edmund Jephcott and Harry Zohn, edited by Michael W. Jennings, 101–133. Cambridge, MA: Harvard University Press, 2006.

Euvrard, Gil-François. "*La presse en Afrique Occidentale Française des origines aux indépendances et conservée à la Bibliothèque Nationale*." Mémoire, École Nationale des Bibliothèques, France, 1982.

Förster, Till. "Visual Presence and Competition in Urban Africa." *Critical Interventions, Journal of African Art History and Visual Culture* 1, no. 2 (2008): 59–79.

Fox Talbot, William Henry. "A Brief Historical Sketch of the Invention of the Art." In *Classic Essays on Photography*, edited by Alan Trachtenberg, 27–36. New Haven, CT: Leete's Island Books, 1980.

Gbadegesin, A. Olubukola. "Photography and Performance in the Yorùbá Photoplay Series." Unpublished paper presented at the Arts Council of the African Studies Association (ACASA) Triennial Symposium on African Art, August 9, 2017.

Haney, Erin. "Film, Charcoal, Time: Contemporaneities in Gold Coast Photographs." *History of Photography* 34, no. 2 (May 2010): 119–133.

Jaji, Tsitsi. *Africa in Stereo: Modernism, Music, and Pan-African Solidarity*. Oxford: Oxford University Press, 2014.

Kasfir, Sidney. "Review of Johannes Fabian, *Moments of Freedom: Anthropology and Popular Culture*." *American Anthropologist* 102, no. 4 (2000): 937–938.

Mbembe, Achille. *On the Postcolony*. Berkeley: University of California Press, 2001.

Newell, Stephanie, and Onookome Okome. *Popular Culture in Africa: The Episteme of the Everyday*. London and New York: Routledge, 2013.

Perret, Thierry. *Le temps des journalistes. L'invention de la presse en Afrique francophone*. Paris: Karthala, 2005.

Picton, John. "Colonial Pretense and African Resistance, or Subversion Subverted: Commemorative Textiles in Sub-Saharan Africa." In *The Short Century: Independence and Liberation Movements in Africa, 1945–1994*, edited by Okwui Enwezor. München: Prestel, 2001.

Pinther, Kerstin. "Textiles and Photography in West Africa." *Critical Interventions* 1, no. 1 (2007): 106–118.

Quayson, Ato. "Signs of the Times: Discourse Ecologies and Street Life on Oxford St., Accra." *City & Society* 22 (2010): 72–96.

Rabine, Leslie W. "Remapping the Worldview of Fashion." *International Journal of Fashion Studies* 3, no. 2 (2016): 185–208.

2 Writing about sound

The early talkie film periodicals of India

Olympia Bhatt

The earliest writings about sound as part of the new talkie films in India present new possibilities in rethinking how sound and questions about aurality are internalized in the emerging cinematic discourse. The two decades that followed the first Lumière film screening in Bombay,[1] as part of an itinerant film exhibition, mark a major period of growth of the film industry, with an increase in both film production and exhibition. Further, the rise of periodicals specifically dedicated to cinema in the 1920s is another indication of not just the growing presence of "cinema habit," but also a concrete attempt to consolidate the nascent local film industry. Coincidentally, the late 1920s was also a period when synchronized sound was introduced in the country in talkies such as Universal's *Melody of Love*, which was screened in 1928-9 in the cities of Bombay and Calcutta,[2] while big studios like Madan acquired RCA sound recording systems to produce short sound films, a vignette of a variety of performances.[3]

This period and context is an interesting moment to look to in furthering ongoing debates about orality/aurality as a source of knowledge, often marginalized in the disciplinary methods of anthropology and ethnography. This disciplinary oversight has led to contemporary revisions and reexaminations of existing methodologies. Veit Erlmann's introductory essay in *Hearing Cultures: Essays on Sound, Listening, and Modernity* identifies how the definition of culture as "acts of inscription" deciphered through "reading and interpretation"[4] has been the main reason for the relegation of all other perceptual, sensory sources of knowledge. He further adds that this second sense is not sequestered or separate from other senses but seems to "have worked in complicity with the panopticon, perspectivism, commodity aesthetics and all the other key visual practices of the modern era we now know so much about."[5] Some of the essays in his compendium explore these existing relationships in creative ways, so as to underscore the often overlooked sonic regimes that exist in the archive and provide new ways of engaging with the subject. For instance, Bruce R. Smith's essay on acoustic archeology of early modern England not only describes the existing historical objects used for creating sounds,[6] but also looks at how a variety of sonic inflections are deducted from the typography of texts of this period. Emily Thompson discusses "the cultural meanings of the new technology to those who deployed it,"[7] referring to the earliest batch of ERPI sound engineers who traveled all over the world to install and maintain the earliest sound equipment. She maps the shifts in the earliest expectations of these engineers and their training, which emphasized and reinforced a certain kind of listening in different spaces and amongst diverse audiences in different parts of the world. According to Thompson, after the initial euphoria around the new technology, sound acquired "unique and competing national characters"[8] in different regions, which, at times, were at odds with the earliest expectations of these technicians, or with what these new sonic technologies purportedly represented.

Aurality in the time of talkies: films, periodicals, and the late colonial period

The focus of *Hearing Cultures* is on the cultural significations and social relevance of important aural functions, listening and hearing, in the existing disciplinary method. These sonic encounters and negotiations perpetuate some of the existing cultural divides, which distinguish "rational" listening as desired and shaped by modern sonic technologies from "primitive" modes of listening. According to Erlmann, this categorization is far more nuanced and can be described as a complex dialectical process "insecurely poised between the modern and the 'primitive,' between the rational and the affective, the discursive and the embodied."[9] This negotiation between the old and the new kinds of aural functions is a common phenomenon in the early twentieth century across the world, where the difference is in how these two manifest themselves and can be approached for a given historical moment and place. In this chapter, the site of this ongoing negotiation are the film periodicals contemporary to the advent of synchronized film sound in India. Within the rapidly evolving soundscape of early twentieth-century India, the film magazines have an added significance beyond the already existing symptomatic relationship between magazines as part of the industrial infrastructure created to supplement the commodity culture around cinema. The advent of talkies is an interesting moment in the development of the sonic landscape in India as new synchronized sound technologies were another addition to the ongoing reconstitution of the existing soundscape. The gramophone was already a popular form, while the radio's introduction preceded talkies' arrival by a few years, amongst a host of other aural technologies such as the public address system, microphone, and telephone, whose presence and usage remains to be documented in the Indian context. What makes the sound of talkies an interesting area to map sonically is the concurrent rise of discourse in the periodicals, exclusively dedicated to cinema, that become popular during this period.

No doubt, the rise and popularity of film periodicals coincides with the high noon of the studio system in the US and proliferating film industries in other parts of the world, including India. These film magazines supported and commoditized star systems and other aspects of cinema through a new kind of visual engagement. Not only personal stories, interviews, and narratives about films and their actors abounded in the magazines of this period, but these periodicals also engaged in a surplus visual economy by providing reproductions of movie stills or nonfilmic shots of film stars, such as close-up portraits, to bolster their celebrity narrative. These additional images and pictures in the magazines point to sources of information and knowledge that are not necessarily bound to the structures of institutionalized knowledge. Neepa Majumdar in her study of the earliest talkie stardom in the Bombay film industry mentions that the "very 'being' of film stars was one of ill repute, [as] the public did not need to *read*, though they may have *heard*, unsavory details about a specific star."[10] The "reputation" of stars in India was not solely reliant on information read in film magazines but was more often "heard:" mostly rumors and gossip that circulated about the personal lives of actors and their nonrespectable, dubious origins, as well as about other people from the film fraternity in general. These magazines, interestingly, implicitly acknowledged these aural networks, often laterally mentioned in their writing only to be rejected and dismissed by alternative narratives about the topics in discussion.[11] In no time, this oral manifestation was not the sole aural source for these journals as the newly introduced synchronized film sound also directly influenced the ongoing cinematic debates about its role and presence in the new talkie. As this new sound was bound to new technology from the West, it was assumed to perpetuate a modern rational listening that would stand in stark contrast to indigenous "primitive" modes that defined the local listening competence. The cinematic discourse in these magazines was, without doubt, part of the "dialectical process" where the role of sound and image in film was first being defined, even as the reverberations of existing aural/oral influences and culture were negotiated in the existing cinematic discourse. It is this tension that will be a central focus of this chapter.

Though the new talkie films constitute an important aural archive of the period, the focus here will exclusively be on the film periodicals of the early talkie period, which are part of a larger and extended history of periodicals in the country. Though the Indian periodical history is vast and diverse and is beyond the scope of this chapter, I will outline a brief history of this new kind of writing that emerged in the late eighteenth century in the subcontinent. The earliest form of the periodical emerged as part of colonial administrative survey, where all aspects of local lives were extensively studied and documented to supposedly aid in effective governance of the region. These documents, no doubt, provide a rich, albeit colonial, history of this period. Meanwhile, the locally produced and distributed periodical also became a popular site to articulate questions of social reform, especially about women's status in the nineteenth century. According to Julie F. Codell, these journals provided unique solutions for reform, borrowing from both eastern and western education, political, and social systems. She further adds:

> The periodical press of India debated these larger changes, mirrored in the alternating patterns of British censorship, surveillance, and liberalization of the Indian press and Indian responses to these policies and to Raj governance in general. [. . .] The Indian press run by and for Indians can be divided into vernacular and English-medium, both of which expressed wide-ranging

> views on many topics and were not politically or ideologically monolithic. [. . .] The vernacular press was often motivated by regional and sectarian issues and readership.[12]

Cinema remained largely outside the purview of these periodicals as "emerging national identity"[13] was a subject of greater concern. However, these journals became the literary precursor of the film periodicals of the twentieth century, even as the latter also borrowed from its precedents in other parts of the world. Some of the film periodicals considered in this chapter include *Filmland, Cinema, Filmindia, Mauj Majha* – the last of which is a Gujarati film-based journal – as well as the daily newspapers, which were the initial sites that preceded these magazines exclusively dedicated to cinema. A cautionary note must be added before taking on these materials as primary sources for this chapter. The extant film periodical archive from the earliest moment of synchronized sound is uneven and skewed, and it was only towards the close of the 1930s that the sources become more periodic and frequent. Second, the ideological leanings determine how certain events are discussed and presented in each of these sources. It also indicates the different networks constituting this new kind of publication. For instance, during my research I found some of the new sound devices mentioned in one publication were unknown in another; the market and publicity of these devices was uneven. Moreover, my limitation in engaging only one kind of linguistic (primarily English) archive assumes to limit the scope of my research for my argument here. However, my intention in this chapter is to initiate a discussion about the topic at hand, which can be further enriched and enhanced through vernacular and other regional archives. Also, the title of the chapter refers to periodicals in India broadly, though most of my sources are publications based in the cities of Bombay and Calcutta (and also Pune, Lahore, with the rare reference to Madras), which were the first significant locations of film exhibition and production in the country; and more important, filmic materiality of this period implicates these two locations. Bombay is an interesting metonym for not just the city but also the larger state of the Bombay Presidency,[14] as well as representative of a desire for a certain kind of cinema.

The ongoing sociopolitical and economic transformations in the 1930s also impact the film industry's growth as well as debates around the new talkie form. The 1920s and 1930s were a period of rapid socioeconomic transformation in Bombay, as the Great Depression loosened the older forms of colonial economic ties, while increasing textile mill productions.[15] In colonial India, the economic slowdown is observed in the agricultural sector, which led to the transference of human and economic resources to different local industries. One such industry to benefit from this shift was the local film industry.[16] This shift also caused the rapid transformation of city spaces, as the rural agrarian slowdown meant that cities became the center of growth as migration of labor, resources, and popular rural forms of entertainment and expression converged with new technologies and machinations of urban life. The rejuvenation that music received in this period – especially in the talkie film – and its bourgeois critique can be situated within this existing trend as the old, feudal economy becomes part of the emerging cultural one.

This period is also marked by a growing sociopolitical divide, aided by the colonial policies that fragment the growing nationalist movement during this period. The colonial government tried sharpening the communal and religious differences through legislative policies such as separate electorates for depressed classes,[17] which brought caste, religious, and other identity issues to the fore. The nationalist movement countered this through different discursive and argumentative strategies,[18] a narrative that also becomes part of the cinematic representational mode to imagine the new nation in the films; it also shows how the local film industry managed to develop and incorporate new technologies in spite of skewed colonial policies hampering its development.[19]

The characteristic, defining feature of sound in Indian talkies was the dominant and overwhelming presence of music, which has been studied in great detail.[20] Continuing on this engagement with music in the early films, I want to draw out the earliest modes of engagement in defining aurality and aural functions of listening and hearing with regards to film music. Music here refers to an evolving format of songs in films with accompanied music, which was initially recorded as live performance but soon became standardized in terms of recording process and duration, the latter linked to the gramophone record.[21] Much early writing on the talkies is concerned with questions of what constitutes appropriate sound for the talking film – the relationship between synchronized sound and the filmic visual; how listening and hearing should be defined for a talkie; the listening competence of the new talkie spectator; the kind of inadvertent, unwanted sounds that become part of the talking film, etc. Music, interestingly, in these discussions features in a paradoxical relationship to the new talking film, as an impediment in the development of an "intelligent," "rational" form of listening/hearing even as it continued in popularity and proliferated as part of the new film form. Its other significant impact is on the visual narrative of stardom as singers and singing became the de rigueur expectation from these new films. Talking films in India suddenly made the singing voice the most desirable feature of the new actor.

This chapter will focus on two sonic facets defining the new talkie in the 1930s: the relationship between music in these new films vis-à-vis the implied questions about noise and other undesirable sounds of the talkies. This association between music and noise is symptomatic of bigger questions: Who is constituting the emerging listening spaces/formations where such categorization is taking place? If the criticism of film music is so widespread, then how did it continue to be so popular and prolific in the films? What do these ongoing discussions about the definition of film sound indicate about the existing sociopolitical dynamics in the country? What role does the new film periodical play as part of these issues?

The second aural feature that will be discussed in detail is how the singing voice is drawn into the existing narrative of film stardom in these film magazines and journals of this early period. The term "singing/singer-actor" defined the skills demanded of the film actors by the new film form.[22] The singing voice also brought an added erotic charge to the visual regime of the star narrative as the actor's embodied presence had its unique set of significations during this period. Though the material evidence specific to the singer-actor Rajkumari remain scarce and limited, her journey as a singing actor from the earliest years of sound is important to document the changes taking place in the first decade of sound recording, and though the singing voice and songs are relegated to the background, their presence and associations continue to seep through in the film magazines of this period.

Noise around film music: the literary precedence to the aural

> Primordially, the production of music has as its function the creation, legitimation, and maintenance of order. Its primary function is not to be sought in aesthetics, which is a modern invention, but in the effectiveness of its participation in social regulation.[23]

The unlikely relationship between music and noise has been theorized in Jacques Attali's *Music: The Political Economy of Noise*, where instead of focusing on the obvious disruptive and undesirable presence of noises, he describes the historical development of music in Europe as a "channelization of noise."[24] Noise literally represents the interference in the reception of a message intended for a receiver, while at the same time it is also engaged in a power struggle with other kinds of sounds for an aural representation of a social order. This ongoing struggle and categorization is somewhat challenged in Paul Hegarty's *Noise/Music: A History* as he discusses how incidental, unwanted sounds, and sometimes even silence, were being incorporated as music in twentieth-century musical experiments. The ongoing tussle between noise and music is literalized in Hegarty's work while challenging the stark difference established between the two in Attali's. Another study that draws these two categories into discussion is Karin Bijsterveld's study of noise and how it shapes arguments about "auditory privacy" in the twentieth century. Her seminal work *Mechanical Sound: Technology, Culture and Public Problems of Noise in the Twentieth Century* raises the perplexing problem of noise and its subjective character in defining hearing, the abstract nature of which makes it a challenge to arrive at a commonly agreed upon definition of the phenomenon. In her case study of the gramophone as a source of noise in the Netherlands in the 1920s and 1930s, she demonstrates how a "reliance on objective scientific standards failed to be productive [in defining]: the case of domestic, or neighborly noise."[25] She further adds that the discussion around the gramophone as a source of noise made it an instrument for the masses and what some defined as "making noise" was more about "sharing music."[26]

In the Indian talkie context, the rhetoric of noise is a complex phenomenon in the discourse of talkie film magazines. In the initial years and early phase of talkie production and distribution, noise is associated, surprisingly, with music and songs of the talkies when the inchoate nascent devices – at different stages of their productive life – were a chief source for this byproduct of acoustic modernity. However, any critique of early sound technology (or any technology for that matter) was complicated by the nationalist discourse, where local indigenous enterprises was celebrated, arguably as an example where Indians were succeeding in spite of the restrictive colonial regime.[27] But what is of more interest here is how the uneven quality of synchronized sound receives mention in these early Indian film journals. The recording or projecting of sound at this time was typically via secondhand and second-rate equipment that was often bought from itinerant travelers and exhibitors visiting the country. The chief concern of technicians was to minimize noise in the audio-track of the film and ensure that these shortcomings did not affect the perception of sound in the film.

Apart from the woes of working with second-rate equipment and outdated technology, the paraphernalia needed for sound recording hints at the sound quality and other challenges in the process of projection or recording. It also indicates the dependencies that facilitate an efficient and audible development of sound and suggests factors that on their own cannot be incorporated into the contemporary politics and its discourse. The audio qualities of talkies or the infrastructural needs to project them, such as a steady supply of electricity to power the sound system, is rarely mentioned in the magazines. The apparatuses featured in Figures 2.1 and 2.2 are examples of such auxiliary devices that directly engage with the concerns

Figure 2.1 DC to AC current converters for talkie machines and "touring talkie outfits."

Reference: *Cinema*, Nov. and Dec., 1932, 52.

Source: National Film Archive of India.

Figure 2.2 "Cineglow": triode vacuum tube for talkie production and exhibition.
Reference: *Filmland*, April 8, 1933, 24.
Source: National Film Archive of India.

of film sound. "Talkie machines, amplifiers, radio-gramophones, radio receivers, touring talkie outfits" are specifically associated with AC generators, Janette DC converters, and other kinds of electricity-producing devices (i.e., dynamo), as seen in Figure 2.1. This device is typically neither associated with cinema nor sound, unlike the triode vacuum tube described in the second figure. The focus of the "Cineglow 3 element recording lamp" is on minimizing distortion and maximizing volume, part of the initial challenges of sound recording.

The noise from technology and the quality of film sound in the initial period rarely finds mention in the film magazines, which reserved most of their critiques for film music. This bias is indicative of how this new genre of writing affiliates to the bourgeois aspirations for the new form. Kaushik Bhaumik argues that the growing popularity and consolidation of cinema in the 1920s made it a subject of bourgeois interest;[28] while others – such as Ira Bhaskar, Richard Allen, and Rosie Thomas – have argued that the bazaar with its Islamicate, Persian, and Oriental influences has been a major determinant of the cinema produced in the same period. What definitely remains an obvious bazaar influence in the talkies and gramophone is the presence of music as the key sound element of these sonic technologies. This overwhelming presence of a popular sonic regime was neither rejected by the new talkie studios nor did the euphoric enthusiasm for it lessen over a period of time. Rather, the treatment of music and sound technology in these magazines is an interesting account of bourgeois negotiation of a popular space whose sonic character is obliquely hinted in different aural narratives.

The striking and common feature of all journals of this period is the large and looming presence of different sound devices. A large number of advertisements and other publicity material is found in the film magazines of this period, featuring a vast number of lesser-known devices that rarely feature in the grand narrative of global film sound history. Figure 2.3 presents one such example, of the Canady sound recording device in the Bombay-based *Cinema* magazine.

The eponymously and appropriately named Chicago Telephone and Radio Company[29] – the distribution company for Canady – was one of the few companies of this period that dealt exclusively with different sound devices since the early twentieth century. Sometimes the films' distribution companies were also directly involved in the publicity of these devices. Figure 2.4 from the Calcutta-based *Filmland* magazine shows how the promotion of sound devices and talkies were in tandem for these film distribution companies.

The "Portable Talkie Set" mentioned in Figure 2.4 indicates the huge market for itinerant exhibition[30] as, initially, the number of film theaters with installed sound projectors were limited. These little-known devices competed regularly with established and expensive devices produced by RCA and Western Electric, which were known and available in the country and also prohibitively expensive. However, it became one way of writing and publicizing sound technology of this period, as seen in Figure 2.5, in an advertisement of the International Service on Sound Corporation in a Gujarati-language film journal called *Mauj Majhu*.

This awareness of the "superiority" of RCA and Western Electric sound projectors presents one version of how film sound circulated in the nascent film industry through the dominant narrative and associations around technology. However, sometimes local predilections and preferences disregarded such claims and deflected it to other concerns. One *Filmland* editorial, for instance, dismisses these claims of technological superiority of the RCA recorder with the following:

> [T]he results of the RCA would have been flawless, if undivided attention were paid to the set by the Indian sound engineers who understand the modulation of sound and the essentials of Indian accent, but unfortunately the affairs in Madans and the conditions of the RCA preclude the possibility of a really efficient Indian sound expert, recording the sound.[31]

The claims made in the editorial are correct about technology needing infrastructural support and technical wherewithal for efficient performance. What is interesting is how the technical proficiency and studio facilities at Madan,[32] one of the most prominent studios of the silent and early talkie period, are dismissed in this analysis. These claims do ring in an iota of truth as Indian silent studio spaces and facilities were sparse and functional in their approach, which was the prevailing

Figure 2.3 Canady sound-on-film recording system.

Reference: *Cinema*, April, 1932, 6.

Source: National Film Archive of India.

Figure 2.4 Film and talkie "Portable Set" distributor, Lucky Films Syndicate, Calcutta.
Reference: *Filmland*, November 19, 1932, 11.
Source: National Film Archive of India.

reality for most Indian studios at the time, including those that survived the sound juggernaut. The same article adds this reality towards its conclusion:

> Thus we find, that only for the "set" Madans stand first in recording, but RCA cannot be taken as a criterion, as it is too costly for an Indian concern to enter the competition with this set. The rest are cheaper sets which are in use in Indian studios.[33]

This critique is significant as it lays the onus on the studio's management of its resources rather than its technological acquisition, while challenging the "superiority" of a monopoly like Madan. It also represents an ongoing social and cultural dynamism at play. Cinema became an important subject of engagement for the Indian middle class, whose will-to-document are part of the "last 100 years [. . .] of a modernity that British colonial rule inaugurated in 19th century India."[34] This middle class in India has been described as a class educated in the newly introduced English education in the nineteenth century and part of the colonial administration and economy. The state of Bengal, where *Filmland* was based, also had a sizeable influential and educated middle class with a growing interest in cinema. Bhaumik claims that interest peeked in tandem with new technology being introduced that made cinema identifiable as a modern form.[35] In such a scenario, Madan studios acquiring the latest RCA recorder remains an anachronism as it represented an old, populist approach to talkie production where popular elements from Parsi theater, which is also what the

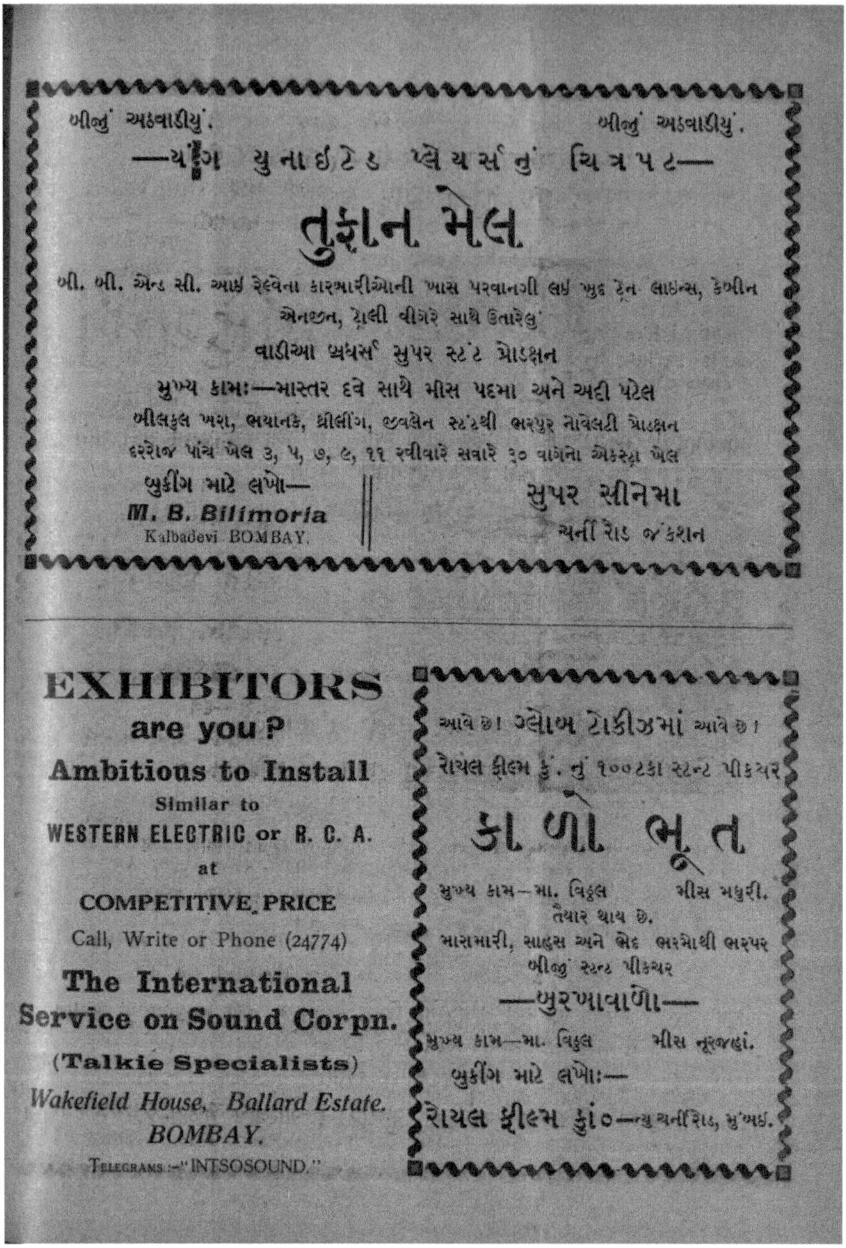

Figure 2.5 Superior and "Similar to Western Electric or RCA" sound projectors for theaters, from International Service on Sound Corporation, Bombay.

Reference: *Mauj Majha*, July 10, 1932, 14.

Source: National Film Archive of India.

Madans were associated with, allowed for an overlapping identification for the audience. This approach would negate any possibility of a development of a modern form even though the best sound recording device was available. This underlying issue is possibly hinted at in this editorial: the old ways of filmmaking being incompatible with new technology and its potential for new forms.

Another version emerges in the Bombay-based journal *Cinema* and its approach to new sound technology.[36] This journal was not merely about the articulation of opinion of a particular class but was driven and funded by the film industry as professional and viable in the colonial economy. This motivation was partly because the feudal economy of certain parts of North and Western (colonial) India had ventured into investing in the film industry, and a journal was helpful in circulating information and developments in the studios of this period as well as publicizing technologies, films, and technical agencies and experts. It made sites such as Lahore, a city in present-day Pakistan, a popular location for artists, producers, and financiers, as well as a source of tremendous influence on the Bombay film industry,[37] something which *Cinema* also documents in its variety of articles and also in the predilections of its filmmaker-cum-editor,[38] B. R. Oberai. What is interesting is how technology was gauged and presented in such a context and what it meant for the industry. Technology aiding in the popularization of existing feudal modes of entertainment such as courtesan music, and performance remained a viable means of continual and new ways of access to older forms. A *Cinema* editorial directly mentions the devices acquired by different studios, though unlike *Filmland*, it remained less critical of the local film industry's efforts of sound conversion:

> Madan Theatres Limited were the first to have RCA Recording System in India and their films were quite popular; now in Calcutta East India Film Co. has another RCA recording set and with the cooperation of good writers, artists and technicians. We may hope they will be able to give India better talking pictures, than hitherto produced by many producers.
>
> Radha Film Company has already got a superior kind of recording set. "Shikari" a talking film produced in India by Eastern Films Limited recorded on a similar set and has excellent recording and we hope Radha Film Company will also produce films with superior recording.[39]

The variety of devices for production only shows the possible variety of sounds across the film studios. As much as it is interesting to understand what the audiences were actually listening to in this period, certain trends are also observed for the initial years of conversion. The impact of talkies was huge on the existing silent film industry even when talkie production was in its nascent stage.[40] Even as early as 1931, when the total number of talkie features produced in the country was 24, distribution companies such as Famous Pictures directly acknowledged their significance in its publicity of silent films with slogans like "silence is golden" and the "film is the language of images" while promoting its recent silent pictures (Figure 2.6).

This overshadowing of the prevailing film culture by the talkie craze even in the absence of a large number of local talkies could mean any of the following: the theaters and itinerant modes of film exhibition in the country had made the

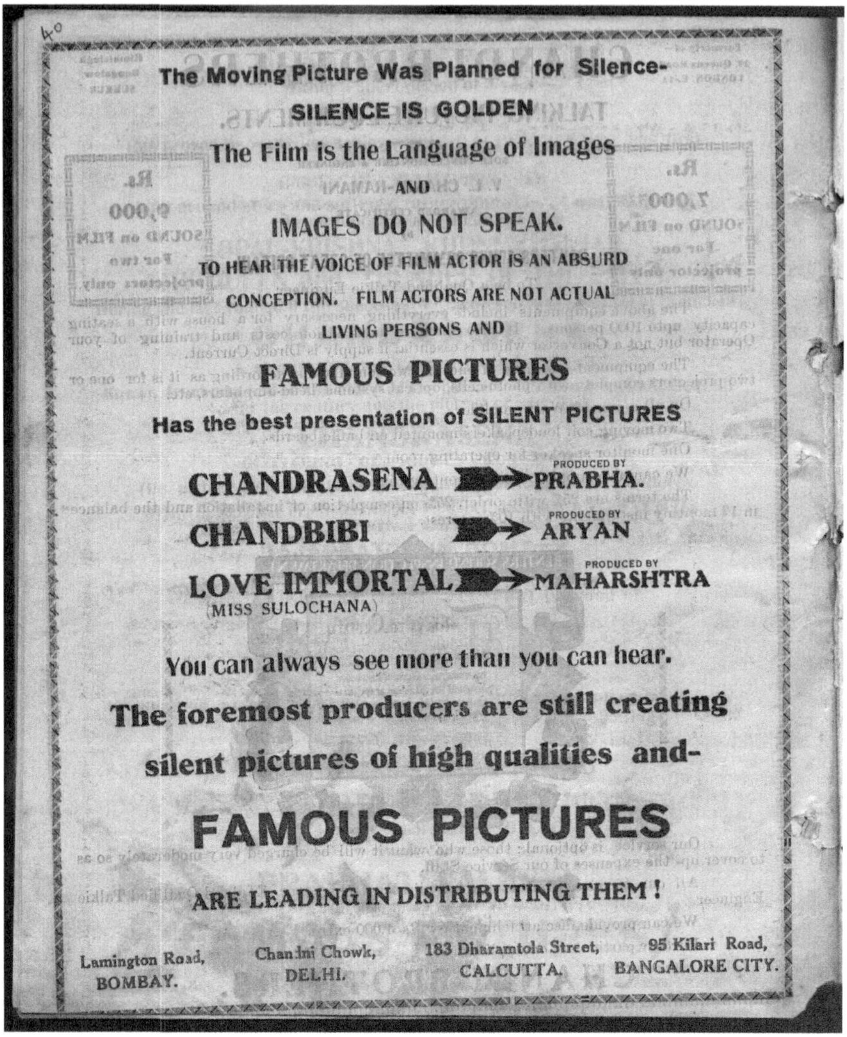

Figure 2.6 A “loud” defense for silent films: distributor Famous Pictures and its latest silent Indian films.

Reference: *Cinema*, May 1931, 40.

Source: National Film Archive of India.

exhibition of this new film a lot easier and more popular; or at least, talkie shorts or partial-talkies were easier to produce and to exhibit, which possibly added to this initial popularity. The language used to describe talkies of this transitional period, in the absence of available films, hints at the latter trend. Many popular silent films directly mention how early talkies’ sound was quantified and referenced, as in the example of Figure 2.7, in which an advertisement describes the popular

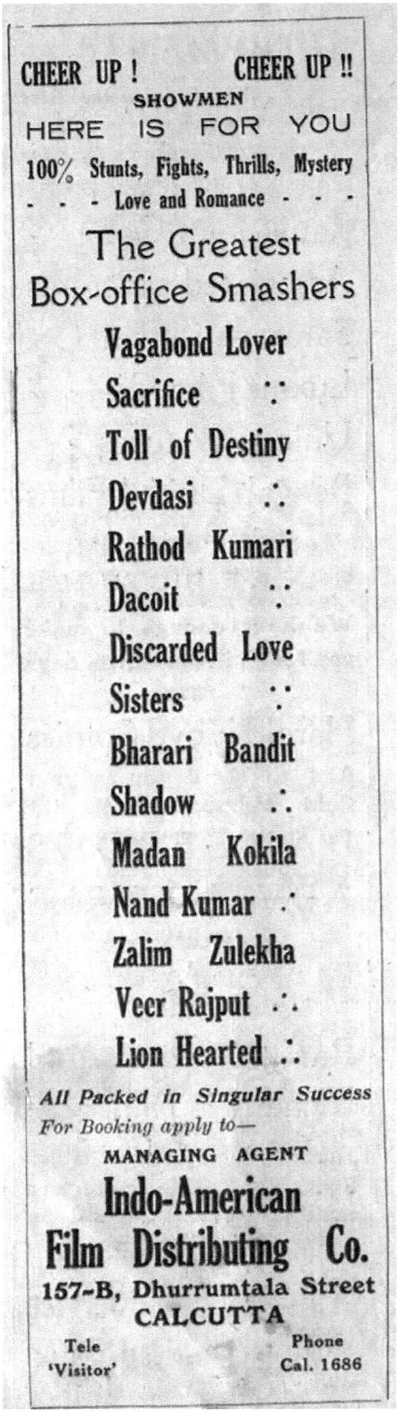

Figure 2.7 Indo-American Distributing Company, Calcutta, and its latest silent films.

Reference: *Filmland*, August 13, 1932, 8.

Source: National Film Archive of India.

Figure 2.8 Lahore-based distributor for RCA Photophone, Empire Projection and Sound Service.

Reference: *Cinema*, August 1933, 21.

Source: National Film Archive of India.

silent genres as "100% Stunts, Fights, Thrills, Mystery, Love, Romance." This is meant to counter the convention of "enumerating" sound as "100%" talkie, a recurring feature of its early publicity. Often, the new talkie studios that emerged insinuated the connection to new sound technology in its name. The Lahore-based studio Playart Phototone Corporation – "The Emblem of Perfect Talkies" – has a similar sounding name to RCA Photophone, to the extent that the former even also imitated the latter's iconic logo. (Figures 2.8 and 2.9).

Further, the talkie studios constantly promoted their newfound sonic credentials in their movie publicity, which makes obvious some of the issues engendered by film sound. Continuing with the Lahore-based studio Playart, the advertisement shown in Figure 2.10 represents its publicity in *Cinema* for the studio's 1932 talkie film *Heer Ranjha*. The aural characteristics of the film are wedged between the film's publicity stills of the images of the lead actors. Based on a well-known romance of Punjabi folklore, the film is described as a "100% talking, singing picturization" whose lead actor, Rafiq Gaznavi, is mentioned as a "gramophone record artist." His education credentials "B. A." is also referenced besides his name and he is given more lines of text when compared to Anwari Begum, the other actor in the film, described merely as the "Nightingale of Punjab," and only named as "Miss Anvar," albeit in large, capital letters. Moreover, "the American expert" William Faulkner is credited for recording the sound on a Rico sound system, which supposedly is a validation of the quality of the film's sound. The credentials of the musicians fits in the existing paradigm ("gramophone record artist"

Figure 2.9 The logo of Lahore-based studio Playart Phototone Corporation and its affinity to (RCA) Sound.

Reference: *Cinema*, April 1932, 12.

Source: National Film Archive of India.

Figure 2.10 Playart Phototone's 1932 talkie film *Heer Ranjha*.

Reference: Cinema, July 1932.

Source: National Film Archive of India.

and "nightingale") while the other reference seems a little misplaced – the reference to Gaznavi's educational background. This kind of information is one way of establishing the film industry's credentials as part of the emerging, educated middle class, employed also in the silent period of cinema, that constantly had to defend its "respectability" by highlighting the participation of the educated class in its development; such advertisement is meant to signal its distance from the ambivalence and liminal space of the bazaar[41] and its networks that nevertheless engendered cinema and other popular forms in the country. The ambiguous nature of the bazaar made it a source of dubious morality, sans respectability, and its influence on silent and then sound cinema furthered these associations.

These respectability debates were further aggravated with the advent of film sound as now the popular sounds of the bazaar resonated with this new technology, too. Popular musicians, singers, and performers became the initial singer-actors for this new form, even as its presence was debated within filmic communities. One of the assumptions of the introduction of new sound technology, especially by the bourgeoisie, was that its presence automatically would lead to the development of an "intelligent" talkie form. This term "intelligent" was imagined as part of a realistic framework of storytelling, whose prototype was the dialogue-heavy synchronized sound of American talkies, and could be used to develop a filmic equivalent of literary realism.

The constant critique of the new talkies was described in one account as follows:

> In the early days of the talkies the outlook was a bit bright. *Everybody expected that intelligent pictures would soon come into their own.* But the success of *Shirin Farhad* caused not a little tremor. Audiences wanted songs and songs. Songs used to take up nearly half the length of the picture. This climax was reached with *Indrasabha* which had more than sixty songs.[42]

Below is another example that invokes a similar term to describe these new talkies:

> The home product enjoys an overwhelming advantage in respect of intelligibility [. . .] dialogue and songs are in languages familiar to the cinema going public.[43]

This concept of the "intelligent" film was engendered by the mimetic nature of sound that was being developed by the Hollywood talking film through naturalized sound. This naturalized approach to sound reproduction allowed for the introduction of literary realism that could now be aurally manifested beyond the "stagey melodrama, the look of performers, anachronisms of décor and dress, Islamicate elements in films and the on-screen female sexuality of popular cinema."[44]

However, silent-period filmmakers transitioning to sound such as Ardeshir Irani of Imperial Movietone were inspired by the Hollywood musical genre for the first talkie feature of India, *Alam Ara*.[45] The success of the musical format made music and songs integral to the new form and even bourgeois-run studios – like New Theatres of Calcutta, Prabhat Studios of Pune, and Bombay Talkies of

Bombay – succumbed to the pressures of integrating music into their films. One of the ways that these new studios distinguished their use of music and songs was by constantly invoking their lineage to the already existent public spheres engendered with the coming of the middle class in India. Theater was an important source of actors (who were well-known theater stars) and some of the storytelling techniques employed in early cinema all over the world. However, in the Indian context its populist association and melodramatic format made it unsuitable for the desired new kind of storytelling. The new studios could not do away with music in their productions but invoked the existing literary sphere as part of their cinematic lineage. Interestingly, all the centers where cinema flourished – cities like Bombay, Pune, Calcutta, and Madras – had a vibrant literary culture that allowed these studios to invoke their realist leanings over populist ones.

Literary realism and its incorporation in cinema was considered a way of introducing relevant contemporary stories and of merging together the literary sphere with a mass medium to improve both the formal and aesthetic values of cinema. Even though silent studios that converted to sound continued to employ the services of stage writers for the films, newly established studios like New Theatres actively sought to use realism-driven vernacular fiction for their films. The studio achieved success with the audience with the adaptation of Sarat Chandra Chattopadhyay's novel *Devdas* (a novel published in 1917 about the contemporary Bengali milieu, while the film versions were released in 1935 in Bengali and in Hindi in 1936). Some studios, like the Imperial Film Company also employed new writers like Sadat Hassan Manto, who is credited as the writer for the film *Kisan Kanya* (1937), and who went on to script other films, including *Apni Nagariya* (1940) and *Naukar* (1945). Studios like Bombay Talkies employed writers like Saradendu Bannerjee and Niranjan Pal, while Marathi writers like Bhaskar Rao, Shivram Vashikar, and P. K. Atre were working with Pune-based studios like Huns Pictures, Navyug Chitrapat, and Prabhat. Filmmakers also tried to adapt stories of popular writers; Premchand's novel *Sevasadan* was made into a film by Mahalakshmi Cinetone under the direction of Nanubhai B. Vakil. Sagar Movietone also adapted the works of well-known Gujarati novelist K. M. Munshi, and Prakash Pictures purportedly made Ramanlal Desai's novel *Poornima* into a film in 1938, starring Sardar Akhtar and Noor Jehan and directed by Balwant Bhatt.[46]

According to a 1940 *Filmindia* article, the film scriptwriter could also evolve from a pan-chewing *munshi*.[47] The article attributes the origins of a literary approach to film writing to Devaki Bose (of New Theatres) and Marathi writer V. S. Khandekar (of Huns Pictures).[48] Interestingly, the whole article does not once mention the presence of songs as an aural element in film; rather it focuses on film technique, and suggests that Hollywood offers strong examples of turning the scripted page into a film scene. According to *Filmindia*, the Hollywood talkie had moved beyond mere adventure films and Westerns – references that were part of the syncretic hybridity of popular cinema.

In spite of these literary predilections and the engagement of the new talking films with the literary sphere, the dominance of songs in films continued without much challenge. Writer Nagara Rao claimed that Indian films should be called "singies" for the undue attention given to songs and music in the film.[49] Any attempt to make films without songs failed miserably. Wadia Movietone released the first talking film without songs, *Naujawan* (1937), a thriller stunt drama directed by Aspi Irani whose plot, according to Hamraaz, did not require any songs as part of its narrative. The film was scripted by J. B. H. Wadia and set an example of Indian cinema without songs. However, a review of the film was not encouraging, as the audience apparently felt cheated without songs, and the film turned out to be a flop.[50] Perhaps for this reason, according to Bhaumik, the new "literary" films typically did not shun away from the inclusion of songs; rather the problem was the song's "disruptive" role in filmic storytelling:

> The Hindu bourgeoisie was seeking to replace Islamicate melodramatic entertainment with its own version of melodramatic realism. If the monumental ethnoscape for India was about "the need to translate Mahabharat stories into ultra-modern ideas," popular cinema was certainly not the right vehicle for it. Dances, Islamicate costumes and songs at the wrong places interrupted the monumentalism of this cinema – be it historical or social realist.[51]

This random inclusion of songs was not only criticized for the heterogeneity and cosmopolitanism of the bazaar influence on cinema but also for internally imbibing these features in its formal development. The use of different kinds of tunes and musical instruments, as part of its background music or orchestra for songs, indicates how songs with their music and their lyrics borrowed easily from other existing musical and poetic traditions. Again in *Filmindia*, this diversity of song is illustrated in an issue from the summer of 1944:

> It has almost become a familiar experience to hear Honolulu calling with all its Hawaiian melodies in the midst of Pilu Thumri sung by some Anil Biswas artiste. We are also acquainted with the Arabic rhythm found in Pancholi pictures given by Ghulam Haider. This type of mixing is all wrong. [. . .] At all times, film music to be popular requires the medium of a melodious voice. The tunes have to be in tune with the psychology of the situations. The words of the song must portray the drama of the situation. And in addition to all this, we must always get a new tune every time.[52]

Even film publicity highlighted this feature of film songs, as in the 1943 advertisement for the film *Chowringhee*, included in the film magazines of the time, and which proclaimed the film to have "rare Egyptian, Arabic, Turkish, and Spanish tunes blended to suit the Indian screen."[53]

Sometimes, these musical borrowings and adaptations were taken to plagiarized extremes, with gramophone companies, too, partaking in the practice of rerecording the popular hits of a particular talkie with slightly modified lyrics or tune:

> No sooner does one [gramophone record] company put on the market a popular song, the other company gets its own stock singer to sing a similar tune

> with almost similar words and very soon the counterfeit record comes before the public to confuse the buyers and disgust the music lovers. Very often one company gets the better of the other by securing the tune and the words even before the release of the original song.[54]

Another justification for the presence of film songs implicitly attributed it to the screening practices in the country. Imported films were usually accompanied with other kinds of short films while the Indian films incorporated the variety film format, which gave it a longer running time than the usual imported film. In *Filmindia*, this phenomenon is explained:

> One reason why Indian films are lengthy is that, having nothing to show side by side, with the feature films, the exhibitors are said to insist on films that last about two and half hours. In the theatres which show English or American films, nearly half the time is generally taken by some short film, some travelogue, a news-reel and often some educational feature in addition. The principal film scarcely exceeds a maximum footage of 10000 [feet].[55]

The next section continues this discussion about music and its impact on the narrative of film stardom. Singers became the new onscreen actors in the talkie era in India as the practice of song recording evolved and was standardized over the decade. Although this decade marks the popularity of the singer, onscreen singing was riddled with ambivalence in the film journals of this period because of their provenance from bazaar networks and the surplus nature of music and songs which, at times, overwhelmed the existing star narrative. I will outline the aural stardom in the career of one of the earliest playback singers of the Indian film industry, Rajkumari, as an aural counterpoint that illustrates the changes to and status of songs and singer-actors in the 1930s. Here, I also argue for how singing is reinvented in the process of filmic self-representation.

Song shaping image: singer-actor shaping the star discourse

> The romance of hearing your own voice on record is something I can't describe.[56]

The narrative of stardom has largely been dominated by the visuality of cinema. Aurality is a rare feature of film stardom that finds little mention in the media texts that construct the polysemy of the star. In the Western discourse of stardom, music and songs feature in the discussion of musicals where the issue of the authenticity of the singing voice and the actor being the same person made musicality an extension of the star image. In the context of Indian cinema, aural stardom is synonymous with the singing voice of Lata Mangeshkar who dominated playback singing for close to five decades.[57] The emergence of Mangeshkar as the top-notch playback singer was a culmination of musical tendencies and discourses that began in this period. Some of these tendencies added a newfound dynamism to the playback song and its visual representation. Rather than focus on the technique of playback and technology of recording here, I will map the impact of aural stardom of the 1930s singing star Rajkumari on the nascent star discourse of the film magazines of this period. Her film acting and singing career present an interesting trajectory where she remained popular as a singer though her onscreen filmic presence was burdened with the constant need, as she self-confessedly put it, to "watch her weight."[58]

Rajkumari Benaraswali[59]/Dubey (1921–2000) was born in the North Indian city of Benaras. She got her first role as a child artiste in the Lahore-based Kamala Movietone's *Raadhe Shyam* (1932). She became associated with Prakash Pictures in the mid-1930s; she claimed to have no musical training, and her singing was restricted to films. Her background song "Nazariya ki maari mari mori guiyyan" from the film *Pakeezah* (1972) became an aural representative of the *kotha*. Her voice was thinner and somewhat comparable to the singing voice of the playback star of the 1950s and 1960s, Lata Mangeshkar. But like all the early playback singers, her voice became a precursor leading to Mangeshkar's success in the decades ahead. In the heyday of Lata Mangeshkar, Rajkumari's voice was used to offset and vocalize the peripheral voices of supporting characters in film. For instance, in the film *Anhonee* (1952), in which Nargis had a double role, Lata Mangeshkar and Rajkumari sang for one of the two twin characters for the song "Zindagi badli."[60] She is one of the earliest singer-actors who survived all the different kinds and forms of sound recording of the 1930s and deliberately chose to focus on her playback-singing career in the following decade.

The evolution of song recording in this decade underwent three different modes of production. In the first stage, when synchronized sound recording was introduced, the film song was recorded as a live performance. This method of recording made theater singer-actors viable and popular performers as their loud voices could reach the single microphone used for recording both the singer and accompanying musicians. This also meant that the film remained the only means of experiencing these performances which were not available in any other sonic format. The preexisting reputation of this new singing actor – from theater and other bazaar networks, gramophone records or nascent radio publicity, or performances – made it easier for film studios to draw the audience to the theaters solely on the reputation of their singing stars. As in the previous *Heer Ranjha* poster where the lead actor is credited as a gramophone record artist (Figure 2.10), Figure 2.11 is a publicity ad for the film *Dharamveer* (1937) whose singer-actors Ratnaprabha and Indira Wadekar are credited as "the melody angel" and "famous radio star," respectively.

At any given moment in the late 1930s to early 1940s there was more than one actor working with the same name. For instance, there were three actors called Leela (Leela Desai of New Theatres; Leela Chandragiri of Saraswati Cinetone, aka "the nightingale of Maharashtra"; and Leela Chitnis of Bombay Theatres), differently spelt Gohars (Gohur Mamajiwala, Gauharbai Karnataki), and so on, which endowed their shared names a worth, a polysemic presence, and a multivalency that was defined and distinguished in terms of their singing voice and screen presence. Some of the new actors shared their name with existing stars in other

Figure 2.11 Sonic reputation resonating in Huns Pictures' 1937 talkie *Dharamveer*.
Reference: *Filmindia*, January 1937, 20.
Source: National Film Archive of India.

genres, and used the name's recognizability to consolidate their presence in this decade, as their familiar name and reputation was far more established than that of the new production companies and studios that had hired them.[61] The growing freelance scenario, with very few studios at this time, allowed stars more flexibility to act in films produced by different studios, while adding to the confusion about the actual identity of the actor in a film.

Even Rajkumari had other eponymous actors, which makes it difficult, in the absence of continuous textual or aural material evidences from this period, to identify and separate out their overlapping histories. There were four actresses with the name Rajkumari in this period: Rajkumari (the name credited in early Madan talkies); T. R. Rajkumari, who featured in the Hindi version of *Devdas* (1937) and followed it with Tamil films in the subsequent decade; Rajkumari Shukla, the actress credited in the Bombay Talkies' film *Jhoola*; and Rajkumari Benaraswali, the singer-actor from Benaras that this section is examining in close detail.

The soaring popularity of songs and music in the talkie films results from the multivalent presence of film songs for aural consumption. The practice of making gramophone records of film songs, which became a standard practice around 1934, made playback as a song-recording technique possible. In this technique, the prerecorded gramophone record was played while the visual was recorded and the actors lip-synced to the recorded song. Initially, the same actors sang as well as performed in the film, a trend that would be discontinued as there was a segregation of the acting and singing labors of the singing star. The success of the film *Khazanchi* (1941) changed the perception of songs, singer-actors, singers, and actors as the success of its music became central to the film's identification. On the cover page of the film magazine *Filmindia*, the filmic picturization of the song "*Sawaan ke nazaare hai*" (which is also written in Hindi here) is directly referenced, with additional sonic resonance as the film's soundtrack gramophone record replaces the wheels of the bicycles of the two lead actors (Figure 2.12). This film is a landmark as it made playback's segregation of the actor and singer an open secret. Initially, the focus remained on the songs of the film and its popularity amongst people from all walks of life, as seen in Figure 2.13, which pictures in cartoon graphics people going about their daily lives and singing, and is accompanied by the text, "our artist's impressions of the popularity of (the song) '*sawan ke nazaare hai.*'"

However, the success of this film also made questions about the practice of playback hard to ignore. A 1943 *Filmindia* article titled "Ghost voices of the screen" described this segregation resulting from playback. Interestingly, the question was neither about the singing voice embodied in the presence of the film actor and the question of authenticity, nor was it about deception inherent in this new approach. The focus, above all, was how playback had resulted in the mechanization of the human voice through an industrial practice where the same set of playback singers – Rajkumari and Amirbai described as "professional ghost singers of the screen" – sang for a large number of actors on the screen. The demand

Figure 2.12 All about song and sound: November 1941 *Filmindia* cover, an ode to the popularity of *Khazanchi*'s music.

Reference: *Filmindia*, November 1941 cover.

Source: National Film Archive of India.

Figure 2.13 Singalong song of *Khazanchi*: "*Sawan ke nazare hai.*"

Reference: *Filmindia*, November 1941, 20.

Source: National Film Archive of India.

instead was for a range of different singing voices to complement the equally diverse number of actors on screen and the affective engagement with the singing voice that occurs in the process of identification:

> They [Rajkumari and Amirbai] have been repeated too often and whosoever's the face, experienced film-goers spot the voice as belonging to one of these two. Once the identification has been done where is the emotional thrill in the music?[62]

And yet, despite the playback tactic being well known, it is the screen actor who would continue to be promoted in film advertisements as the singer. After the success of *Khazanchi*, its star Manorama was described in the Caravan Pictures' film *Bhai* as follows: "Here comes the nightingale Manorama with melody on her lips, rhythm on her feet, singing and dancing her way into your hearts" (Figure 2.14). Singers Naseem Akhtar and Zeenat Begum are credited as the singers for the film though there is no reference to them in the film publicity.[63]

The other dichotomy that the singing body of Rajkumari represented was the incompatible filmic presence on screen due to the physical expectations placed on stars versus the popular allure of her singing voice. The constant scrutiny of the star body determined the course of her career as a playback singer.[64] What is interesting, however, is that this relationship between the "acting body" versus the "singing voice" is not a simplified binary that defined and pitted one against the other. In an interview, she presents a rather contradictory relationship to her training as a singer, and tries to establish her vocal provenance and influence upon cinema itself.

Her confession to being inspired to sing by film songs and the absence of classical musical training, which would have put her in the popular songstress history, is an interesting means to describe her vocal presence in cinema:

> I was born in a family which did not have much to do with music or singing. But just opposite the place where I lived stood a cinema hall. Close enough for me to be able to hear the songs – film songs, which gave me immense pleasure. I was only six or seven years old then [. . . and would] spend hours with the cinema projectionist who I'd befriended, and who'd allow me to watch the film, standing on a stool in the projection room. And my joy knew no bounds when a song flashed upon the screen and I would sing along with it. There were many famous singers then, but I never got to know them for my *social background was a different one.*[65]

This self-awareness of one's social position as a popular singer versus being a singer for talkies puts her acting and playback career in the context of emerging practices and means of production. What is interesting is this new kind of self-representation through song or music that is not necessarily about access to an interiority that is central to the star discourse;[66] rather it is about how film music is being used to construct an alternate space for self-representation outside of

Figure 2.14 The singer-actor replaced by the onscreen actor: Manorama credited as the "nightingale" for *Bhai* (1944).

Reference: *Filmindia*, August 1944, 45.

Source: National Film Archive of India.

existing older popular forms. And this is part of the self-fashioning that the earliest gramophone singers engaged in in the two decades prior to film sound.[67]

Conclusion

The growing presence of the film periodicals in India logically coincides with the rise of new film technology, as the former was needed to promote and publish the presence of these new devices. And the periodical had to also acquire a discursive approach to talk about sound as part of the new audiovisual cinematic phenomenon. This chapter examined this tussle for definition and control over sound and auditory functions, where the periodical presents a sonic archive with "loud" discussions about the right sounds for cinema and the acknowledgment of the elisions and oppositions from the popular side of the story. The presence of the film magazine resulted in an active development of listening cultures around film music as film magazines like *Film Sangeet*, *Madhuri*, and *Sargam ka Safar* in the later decades, emphasized film music in their writing.

The presence of new sound technology was also discussed, though its earliest reference in the journals tended to appear in advertisements in the periodicals of the time. The other emerging discourse of this period in the film magazine was its negotiation of the impact of the singer-actor on the existing star discourse. Here music and the singing voice's discursive evolution coincides with the standardization of the sound recording practice of playback that resulted in the segregation of the onscreen actor from the playback singer. The stardom of one of the earliest singer-actors Rajkumari helps us to understand how singing in films evolved over the decade as singers shed their earlier bazaar lineage and evolved to address the sole needs of the new technological form, which culminated in the aural stardom and dominance of the biggest playback singer in the post-independence Indian film industry, Lata Mangeshkar. This chapter has provided a context for the career of Rajkumari within the early history of talkies in India, traced through a comprehensive consideration of film magazines of the period. And through close readings of the advertisements in these publications, a greater understanding of the distribution and reception of these films is possible, thus underscoring the need to turn to this largely overlooked aspect of the magazine within the realm of periodical studies, in order to tell fuller histories.

Notes

1 "The agents of the Lumière Brothers brought Bombay's first cinematograph show to Watson's Hotel, at the junction of the Fort and Colaba. Cinema made its debut in Bombay on 7 July, 1897, and soon became a regular feature of the city's landscape." [Kaushik Bhaumik, "The Emergence of the Bombay Film Industry, 1913–1936" (PhD diss., University of Oxford, 2001), 19.]

2 "In December 1928, Universal's film titled *Melody of Love* (A B Heath) was released by Madan Theatres at Elphinstone Palace in Kolkata and was shown on 21 February at the Excelsior Theatre, Mumbai." [Virchand Dharamsey, "The Advent of Sound in Indian Cinema: Theatre, Orientalism, Action, Magic," *Journal of Moving Image*, no. 9 (2010): 24.]

3 Released on March 14, 1931, the same day when Ardeshir Irani of Imperial Movietone released the first talkie feature *Alam Ara*, it was a "portmanteau film [with] as many as thirty-one items, comprising of dance, music, and dramatic pieces, in Bengali, Hindi and Urdu." [Samik Bandhopadhyay, "Introduction," in *Indian Cinema: Contemporary Perceptions from the Thirties*, ed. Samik Bandhopadhyay (Jamshedpur: Celluloid Chapter, 1993), 1.]

4 Veit Erlmann, *Hearing Cultures: Essays on Sound, Listening, and Modernity*, ed. Veit Erlmann (Oxford and New York: Berg, 2004), 1.

5 Ibid., 5.

6 "In a few cases we have direct physical evidence. For example, musical instruments from the period survive and can still be played (Munrow 1976). Some of the same church bells still hang in some of the same belfries and can still be rung. Some of the same interior spaces still exist, and their acoustic properties can still be experienced." [Bruce R. Smith, "Listening to the Wild Blue Yonder: The Challenges of Acoustic Ecology," in *Hearing Cultures*, 22.]

7 Emily Thompson, "Wiring the World: Acoustical Engineers and the Empire of Sound in the Motion Picture Industry, 1927–1930," in *Hearing Cultures*, 192.

8 Ibid., 208.

9 Erlmann, *Hearing Cultures*, 13.

10 Neepa Majumdar, *Wanted Cultured Ladies Only! Female Stardom and Cinema in India, 1930s–1950s* (Chicago: University of Illinois Press, 2009), 40.

11 Most of these writings were about the disreputable nature of the film industry and suspect morals of its actors, especially those of women. See Majumdar, *Wanted Cultured Ladies Only*; Bhaumik, "Emergence of Bombay Film Industry," for a discussion on female stars and "respectability" debates of the early period.

12 Julie F. Codell, "Introduction: The Nineteenth-Century News from India," *Victorian Periodicals Review* 37, no. 2 (Summer 2004): 107.

13 Ibid., 108.

14 The Bombay Presidency was a colonial administrative unit and included most of modern-day Maharashtra, parts of Gujarat and Karnataka, and Sindh province from present-day Pakistan. Until the 1930s, the Aden peninsula (part of present-day Yemen) and Khuriya Muriya islands (part of present-day Oman) in Middle East Asia were also part of this British-administered Indian state. [Susan Wolcott, "Strikes in Colonial India, 1921–1938," *ILR Review* 61, no. 4 (July 2008): 460.]

15 "India's modern cotton textile industry began in the city of Bombay in the 1850s, and by the 1920s there were mills in many parts of India. The Bombay City industry experienced growth to 1922, but there was a sharp downturn from 1922 to 1932, followed by a gradual recovery after protective tariffs were put in place. While the Bombay City industry shrank, other centers expanded, both within and without the Bombay Presidency." [Wolcott, "Strikes," 460.]

16 "Sugar, cement and paper industries developed rapidly in the 1930s, while Tata Steel was strong enough to do without protection after 1934. [. . .] Indian industry also benefited from the fact that agricultural prices declined much more sharply than industrial, while commercial and rural depression probably led to a transfer of capital from trade, usury and land-purchase to industry." [Sumit Sarkar, *Modern India: 1885–1947* (Chennai: Macmillan Press, 1983), 260.]

17 Ibid., 328–331.

18 "More importantly, 'Hinduism' contained, within itself, the solution to irrationality and social degradation of untouchability." [William Gould, "The U. P. Congress and 'Hindu Unity': Untouchables and the Minority Question in the 1930s," *Modern Asian Studies* 39, no. 4 (October 2005): 848.]

19 The nationalist movement and its leaders in India were not supportive of the local film industry as it was not considered a serious representational form and was believed to be of a dubious moral character. However, the film industry and its technological

innovation, especially for the promotion and development of film sound, constantly framed it within the nationalist discourse. Refer to Joppan George, "The Many Passages of Sound: Indian Talkies in the 1930s," *Bioscope: South Asian Screen Studies* no. 2 (2011): 88–89; Olympia Bhatt, "Mapping the Materiality of Sound: A Cultural History of Sound Technology in 1930s Bombay Cinema" (PhD diss., Jawaharlal Nehru University, 2016), 71–78.

20 In the Indian context, film music has been an important area of study in the discipline of musicology; for instance, the work of Alison Arnold (1991) is a first of its kind in analyzing the music in Hindi film songs along with the studio practices of this period. Previously, writing about music in films remained accessible in the works of early music critics like Bhaskar Chandravarkar and Ashok da Ranade or Indian filmmakers like Ritwik Ghatak and Kumar Shahani. The development of the widespread aural culture of Madras Presidency with the proliferation of sound technologies like radio and gramophone in the 1920s and 1930s is documented in the ethnographic account of Stephen Hughes. A recent edition of Jadavpur University's *Journal of the Moving Image* (2007) focuses on the development of sound in cinema. Another scholar who has worked on one of the early talkie studios, New Theatres, Madhuja Mukherjee actively engages with sound in the 1930s in her *Aural Films, Oral Cultures: Essays on Cinema from the Early Sound Era* (2012), a compendium of the kind of writing sound generated in the journals of this period. Apart from these historical engagements, there is a complete alternate sphere of writing on sound accessed through the travels and circulation of aural technologies like the gramophone, radio, and cassettes or the practice of music composition and orchestra in the Hindi film industry as seen in the work of another musicologist, Gregory Booth. The musical feature of the popular film has generated numerous databases and other important websites that document all the necessary information about the singers, music directors, lyricists, films, and other related material. The biggest and most important anthology of Hindi film songs is Kanpur-based Harminder Singh Hamraaz's *Hindi Film Geet Kosh*, organized in individual volumes, each covering a decade of film songs since the beginning of sound. Another encyclopedic account documenting music directors and singers since the advent of sound is Pankaj Rag's *Dhuno ki Yatra*, which has significant data about this period.

21 The gramophone was the aural medium that allowed an additional text for the talkie cinema to circulate in the public sphere. Initially, the gramophone record had a separate recording from the performance of the song in the film. See Bhatt, "Mapping," 173.

22 Even though there were silent stars like Sabita Devi who continued to be popular and present in the talkies, singing was expected and also tried by some of these stars like Sulochana and Gauhar. I want to focus on how the new singer-actor, for a brief moment in film history, managed to hold sway over the stardom narrative before singing receded into the background as playback recording took control. The monopolistic rise of Lata Mangeshkar in the 1950s and 1960s brought the singer to the fore, though the paradigms to evaluate film songs and playback singers had changed. For further discussion on Lata Mangeshkar, refer to Majumdar, *Wanted Cultured Ladies*, 173–202.

23 Jacques Attali, *Music: The Political Economy of Noise* (Minneapolis: University of Minnesota Press, 2009), 30.

24 Ibid., 26.

25 Karin Bijsterveld, *Mechanical Sound: Technology, Culture and Public Problems of Noise in the Twentieth Century* (Cambridge: The MIT Press: 2008), 159.

26 "Significantly, the council members representing the working class conceptually carved out, so to speak, the right of workers to a 'sound culture' of their own. They did so by referring to the low price of the gramophone, by characterizing it as a 'musical instrument for the masses,' and by creatively redefining 'making noise' into 'sharing music'." [Bijsterveld, *Mechanical Sound*, 167.]

27 See note 19.

28 "The period between 1925 and 1928 had seen a steady expansion in production. [. . .] Productions had become more elaborate in terms of spending on stock, (films were longer and more copies were made), star salaries, sets, technology and the increasing expenses of managing a studio with salaried employees and publicity." [Bhaumik, "Emergence," 90.]

29 http://chicago-radio.net/index.php/about-us.

30 Refer to Sudhir Mahadevan, "The Traffic in Technologies: Early Cinema and Visual Culture in Bengal, 1840–1920" (PhD diss., New York University, 2008), 266–314.

31 Editorial, *Filmland*, July 2, 1932, 2.

32 Ranita Chatterjee documents the history and emergence of Madan Studios of Calcutta from the beginning of twentieth century with J. F. Madan, its proprietor being an agent for many foreigners and who soon ventured into film production: "The First World War was a turning point in film finance. J. F. Madan made huge profits in the war as the official supplier to the British Army. This, along with any profits he may have made through his film exhibition businesses across colonial India, led him to expand his topical film unit and enter into feature length productions. [. . .] The Madans' first feature film was based on the well-known legend of king Harishchandra (the connections with Phalke's case are unmistakable). *Satyawadi Raja Harishchandra* released in Calcutta in 1917. Their next film, *Bilwamangal*, released in 1919, is now credited as the first Bengali film by contemporary historians. [. . .] Thereon the Madans embarked on mass scale production of feature films and virtually dominated silent and early talkie production until the early 1930s, when the company stopped production. It is worth noting here that the Madans were the only vertically integrated company in South Asia at this time, producing, distributing and exhibiting its own films." [Ranita Chatterjee, "Journeys in and Beyond the City: Cinema in Calcutta 1897–1939" (PhD diss., University of Westminster, 2011), 67–68.]

33 Editorial, *Filmland*, July 2, 1932, 2.

34 Dipesh Chakravarty, "Witness to Suffering: Domestic Cruelty and the Birth of the Modern Subject in Bengal," in *Questions of Modernity*, ed. Timothy Mitchell (Minneapolis: University of Minnesota, 2000), 49.

35 Bhaumik, "Emergence," 89.

36 One common ground between these two early period journals *Filmland* and *Cinema* is how they treat local innovation of technology within the common nationalist framework. It is not so much about the success of these devices, but making a case for them in the extremely volatile decade of the 1930s.

37 Ashish Rajadhyaksha and Paul Willemen mention how the agragrian money in Lahore was divested into film projectors and other such devices. Companies like Empire Talkie Distributors emerged with the passing of the Land Act of 1901, which led to a reduction of investment in agriculture and instead in the entertainment industry. [Ashish Rajadhyaksha and Paul Willemen, *The Encyclopedia of Indian Cinema* (New Delhi: Oxford University Press, 1988), 17.]

38 B. R. Oberai was a scenarist for the 1932 film *Pakdam Rakshasa*.

39 Editorial, *Cinema*, November and December, 1932, 11.

40 Silent films were produced in the country until 1934, after which talkie production completely took over in cities like Bombay, Calcutta, Pune, and Lahore.

41 Clifford Geertz defines the economic paradigm of the bazaar, which has resonance for its attempt at cultural engagement: "Neither the rich concreteness or reliable knowledge that the ritualized character of nonmarket economies makes possible, nor the elaborate mechanisms for information generation and transfer upon which industrial ones depend, are found in the bazaar; neither ceremonial distribution nor advertising; neither prescribed exchange partners nor product standardization. The level of ignorance about everything from product quality and going prices to market possibilities and production costs is very high, and much of the way in which the bazaar functions

can be interpreted as an attempt to reduce such ignorance for someone, increase it for someone, or defend someone against it." [Clifford Geertz, "The Bazaar Economy: Information and Search in Peasant Marketing," *The American Economic Review* 68, no. 2 (May 1978): 29.] Kaushik Bhaumik illustrates how this bazaar economics determined the cinema culture that emerged in the cities of Bombay and Lahore: "The cosmopolitan *bazaar* cinema melded into the hectic traffic outside. It fitted into the work schedule of the marketplace. People flowed into and out of the hall following the ebb and tide of the hectic economic and labour schedules in the *bazaar*, as did the characters on the screen. The audience read or heard about larger-than-life adventure heroes and heroines from stories told or published in the *bazaar*, read about thrilling real-life adventures in the newspapers, consumed folk musical and dance performances as well as the *fakir*'s discourse, saw the martial regalia of the princes and aristocrats on the streets, enjoyed the *tawaif*'s dances as well as Parsi and *nautanki* plays and watched Sulochana romances in the *bazaar*." [Bhaumik, "Emergence," 140.]

42 Bandhopadhyay, "Introduction," 12, emphasis added.

43 Author Anon, "Stars of the Indian Screen: How the Film Idols Made Their Names," *Times of India*, Indian Film Industry's Silver Jubilee Supplement, May 5, 1939, 9.

44 Bhaumik, "Emergence," 149.

45 Irani claimed *Showboat* (1927) as the source of inspiration for the musical form of his film *Alam Ara*.

46 Baburao Patel, "Give Our Writers a Square Deal," *Filmindia*, September 1940, 4.

47 This term refers to an ordinary clerk who merely transcribes the ideas given to him by the film or stage director. The absence of original ideas and creativity in writing was what distinguished an ordinary *munshi* from the new scriptwriter or literary author.

48 Patel, "Give," 3.

49 Nagara Rao, "Andhra Talkies," *Filmland*, June 3, 1933, 8.

50 The film is described as follows: "Based on a 24-hour crime cum action film, it seemed that songs were not needed in the film. However, after its screening in some places the audience blamed the producers for defrauding them by not including songs in the film." Original: "24 ghante ki apradh evam maar-dhaar sambandhi ek kahani ko lekar nirmit iss film mei geeto ko shaamil karne ki zaroorat nahi samjhi gayi thi. Parantu pradarshan ke baad kuch sthaano pur nirmaata ko darshako ke iss aarop ka saamna karna pada ki geet shaamil na karke nirmaata ne janta ke saath dhokadhadi ki hai." Harminder Singh Hamraaz, *Hindi Geet Kosh: Encyclopaedia of Hindi Film Songs Vol I* (Kanpur: Satinder Kaur, 1988), 380. [Translation is the my own. -OB].

51 Bhaumik, "Emergence," 177.

52 Baburao Patel, "This Land of Music," *Filmindia*, July 1944, 9.

53 Poster for the film *Chowringhee*, *Filmindia*, September 1943, 3.

54 Judas, "Bombay Calling: Gentlemanly Pact Needed!" *Filmindia*, January 1944, 15.

55 Judas, "Bombay Calling: Plea for Newsreels," *Filmindia*, April 1942, 8.

56 Kanan Devi, *My Homage to All* (New Delhi: Zubaan, 2013), 14.

57 Refer to Majumdar, *Wanted Cultured Ladies*; Shikha Jhingan, "The Female Voice in Hindi Film Songs: Performance, Practices and Circulation" (PhD diss., Jawaharlal Nehru University, 2013), 1–56.

58 Ashok Ranade, "When a Song Paid 50 and Petrol Was 6 Annas a Gallon," in *Cinema Vision India: Challenge of Sound the Indian Talkie Vol II*, ed. Siddharth Kak (1983), 16.

59 The term Benaraswali means a women hailing from Benaras. The Hindi terms Wali/Wala are gender-specific suffixes to attribute characteristics of the term to which it is added. In the case of Rajkumari, it refers to her hometown. This last name was one way of distinguishing singers in the late nineteenth and early twentieth centuries to identify their musical lineage, history, and style. She later married V. K. Dubey, who also hailed from Benaras, and took his last name.

60 Khubchandani, Lata. 2010. "Rajkumari – Interview." Accessed April 1, 2016. http://cineplot.com/rajkumari-interview/. "Zohrabai, Amirbai and Rajkumari: Playback," accessed April 1, 2016. www.womenonrecord.com/music-makers/artists/zohrabai-amirbai-rajkumari.

61 Three different kinds of studios are seen in this period. First, silent studios like Madan that managed to transition to sound but could not survive for long. Second, new bourgeois studios like Prabhat, Bombay Talkies, and New Theatres that created dedicated studio space and facilities along with salaried actors who worked in their films. Most of the studios of this period, however, were new, with some publicizing upcoming releases that were ultimately never completed. The closing of old silent studios allowed a lot of resources for film production to be readily available on the market, which these new studios utilized to develop talkies at a fast pace to make a quick gain.

62 Judas, "Bombay Calling: Ghost Voices of the Screen," *Filmindia*, July 1943, 11.

63 Hamraaz, *Hindi Geet Kosh Vol II*, 145.

64 Neepa Majumdar, in her analysis of Lata Mangeshkar's stardom, focuses on how a hierarchy was implied between the female body and the female voice. The former is available for visual consumption and lends itself more easily to scandalous associations, whereas the latter, as established in the persona of Mangeshkar, could encompass traditional Indian values. [Majumdar, *Wanted Cultured Ladies Only*, 191.]

65 Ranade, "When a Song," 16, emphasis added.

66 "Stars not only bespeak our society's investment in the private as the real, but also often tell us how the private is understood to be the recovery of the natural 'given' of human life, our bodies." [Richard Dyer, "Living Stars," in *The Film Studies Reader*, ed. Joanne Hollows, Peter Hutchings, and Mark Jancovich (London: Bloomsbury Academic, 2000), 132.]

67 See, for instance, the work of Jhingan, "The Female Voice in Hindi Film Songs"; Stephen Hughes, "The 'Music Boom' in Tamil South India: Gramophone, Radio and the Making of Mass Culture," *Historical Journal of Film, Radio and Television* 22, no. 4 (2002): 445–473; Stephen Hughes, "Music in the Age of Mechanical Reproduction: Drama, Gramophone, and the Beginnings of Tamil Cinema," *The Journal of Asian Studies* 66, no. 1 (February 2007): 3–34; Vikram Sampath, *My Name Is Gauhar Jan: The Life and Times of a Musician* (New Delhi: Rupa, 2010); Sarah Rahman Niazi, "Cinema and the Reinvention of the Self: Women Performers in the Bombay Film Industry 1925–47" (MPhil diss., Jawaharlal Nehru University, 2011).

Bibliography

Attali, Jacques. *Music: The Political Economy of Noise*. Minneapolis: University of Minnesota Press, 2009.

Author Anon, *Times of India*, February 6, 1929.

Author Anon, *Times of India*, May 5, 1939.

Bandhopadhyay, Samik. "Introduction." In *Indian Cinema: Contemporary Perceptions from the Thirties*, edited by Samik Bandhopadhyay, 1–16. Jamshedpur: Celluloid Chapter, 1993.

Bhaskar, Ira, and Richard Allen. *Islamicate Cultures of Bombay Cinema*. Delhi: Tulika, 2009.

Bhatt, Olympia. "Mapping the Materiality of Sound: A Cultural History of Sound Technology in 1930s Bombay Cinema." PhD diss., Jawaharlal Nehru University, 2016.

Bhaumik, Kaushik. "The Emergence of the Bombay Film Industry, 1913–1936." PhD diss., University of Oxford, 2001.

Bijsterveld, Karin. *Mechanical Sound: Technology, Culture and Public Problems of Noise in the Twentieth Century*. Cambridge: MIT Press, 2008.

Chakravarty, Dipesh. "Witness to Suffering: Domestic Cruelty and the Birth of the Moder Subject in Bengal." In *Questions of Modernity*, edited by Timothy Mitchell, 49–86. Minneapolis: University of Minnesota, 2000.

Chatterjee, Ranita. "Journeys in and beyond the City: Cinema in Calcutta 1897–1939." PhD diss., University of Westminster, 2011.

Codell, Julie F. "Introduction: The Nineteenth-Century News from India." *Victorian Periodicals Review* 37, no. 2 (Summer 2004): 106–123.

Devi, Kanan. *My Homage to All*. Delhi: Zubaan, 2013.

Dharamsey, Virchand. "The Advent of Sound in Indian Cinema: Theatre, Orientalism, Action, Magic," *Journal of Moving Image*, no. 9 (December 2010): 22–50.

Dyer, Richard. "Living Stars." In *The Film Studies Reader*, edited by Joanne Hollows, Peter Hutchings, and Mark Jancovich, 128–134. London: Bloomsbury Academic, 2000.

"Editorial." *Cinema* 11, November and December 1932.

"Editorial." *Filmland* 2, July 2, 1932.

Erlmann, Veit. "But What of the Ethnographic Ear? Anthropology, Sound and the Senses." In *Hearing Cultures: Essays on Sound, Listening, and Modernity*, edited by Veit Erlmann, 1–20. Oxford and New York: Berg, 2004.

Geertz, Clifford. "The Bazaar Economy: Information and Search in Peasant Marketing." *The American Economic Review* 68, no. 2 (May 1978): 28–32.

George, Joppan. "The Many Passages of Sound: Indian Talkies in the 1930s." *Bioscope: South Asian Screen Studies* no. 2 (2011): 83–98.

Gould, William. "The U. P. Congress and 'Hindu Unity': Untouchables and the Minority Question in the 1930s." *Modern Asian Studies* 39, no. 4 (October 2005): 845–860.

Hamraaz, Harminder Singh. *Hindi Film Geet Kosh: Encyclopaedia of Hindi Film Songs*, Vol. 2. Kanpur: Satinder Kaur, 1984.

Hamraaz, Harminder Singh. *Hindi Geet Kosh: Encyclopaedia of Hindi Film Songs*, Vol. 1. Kanpur: Satinder Kaur, 1988.

Hegarty, Paul. *Music/Noise: A History*. London: Bloomsbury Academic, 2007.

Hughes, Stephen. "The 'Music Boom' in Tamil South India: Gramophone, Radio and the Making of Mass Culture." *Historical Journal of Film, Radio and Television* 22, no. 4 (2002): 445–473.

Hughes, Stephen. "Music in the Age of Mechanical Reproduction: Drama, Gramophone, and the Beginnings of Tamil Cinema." *The Journal of Asian Studies* 66, no. 1 (February 2007): 3–34.

Jhingan, Shikha. "The Female Voice in Hindi Film Songs: Performance, Practices and Circulation." PhD diss., Jawaharlal Nehru University, 2013.

Judas. "Bombay Calling: Gentlemanly Pact Needed!" *Filmindia*, January 1944.

Judas. "Bombay Calling: Ghost Voices of the Screen." *Filmindia*, July 1943.

Judas. "Bombay Calling: Plea for Newsreels." *Filmindia*, April 1942.

Mahadevan, Sudhir. "The Traffic in Technologies: Early Cinema and Visual Culture in Bengal, 1840–1920." PhD diss., New York University, 2008.

Majumdar, Neepa. *Wanted Cultured Ladies Only! Female Stardom and Cinema in India, 1930s–1950s*. Chicago: University of Illinois Press, 2009.

Mukherjee, Madhuja. *New Theatres Ltd.: The Emblem of Art, The Picture of Success*. Pune: National Film Archive of India, 2009.

Patel, Baburao. "Give Our Writers a Square Deal." *Filmindia*, September 1940.

Patel, Baburao. "This Land of Music." *Filmindia*, July 1944.

"Poster for the film *Chowringhee*." *Filmindia* 3, September 1943.

Rajadhyaksha, Ashish, and Paul Willemen. *The Encyclopedia of Indian Cinema.* New Delhi: Oxford University Press, 1988.

Ranade, Ashok. "When a Song Paid 50 and Petrol Was 6 Annas a Gallon." In *Cinema Vision India: Challenge of Sound the Indian Talkie*, Vol. 2, edited by Siddharth Kak, 16–18, 1983.

Rao, Nagara. "Andhra Talkies." *Filmland*, June 3, 1933.

Niazi, Sarah Rahman. "Cinema and the Reinvention of the Self: Women Performers in the Bombay Film Industry 1925–1947." MPhil diss., Jawaharlal Nehru University, 2011.

Sampath, Vikram. *My Name Is Gauhar Jan: The Life and Times of a Musician.* Delhi: Rupa, 2010.

Sarkar, Sumit. *Modern India: 1885–1947*. Chennai: Macmillan Press, 1983.

Thomas, Rosie. *Bombay before Bollywood: Film City Fantasies*. Hyderabad: Orient Black-swan, 2014.

Thompson, Emily. "Wiring the World: Acoustical Engineers and the Empire of Sound in the Motion Picture Industry, 1927–1930." In *Hearing Cultures: Essays on Sound, Listening, and Modernity*, edited by Veit Erlmann, 191–210. Oxford and New York: Berg, 2004.

Smith, Bruce R. "Listening to the Wild Blue Yonder: The Challenges of Acoustic Ecology." In *Hearing Cultures: Essays on Sound, Listening, and Modernity*, edited by Veit Erlmann, 21–42. Oxford and New York: Berg, 2004.

Wolcott, Susan. "Strikes in Colonial India, 1921–1938." *ILR Review* 61, no. 4 (July 2008): 460–484.

3 One-time periodicals, unrealized manifestos

Little magazines and the ethos of failure[1]

Nana Ariel

In 2010, a magazine entitled *02010* was published as a one-time collection of scattered papers in a brown cardboard folder (edited by Hebrew writer Oded Carmeli). Among the eclectic literary works and manifestos appeared the famous portrait of the seven members of space shuttle "Columbia," which tragically disintegrated at the end of its mission in 2003, resulting in the death of the crew – six American astronauts and the Israeli astronaut Ilan Ramon (Figure 3.1). The peculiar, inexplicable choice to include this photo among the texts in the magazine seems at first glance like a pseudo-naive, campy fascination with the heroic, (already) nostalgic, and over-repeated image. The tragic pathos of the shiny orange space suits of the proud men and women against the sky-blue background, with the US and Israeli flags peeping out from both sides, must have evoked the editor's desire to tease its readers with an ironic, iconoclastic act of pseudo-commemoration. Nevertheless, in this chapter I suggest a second, anti-ironic look at this editorial choice, and argue that it not only reflects the editor's self-perception, but also epitomizes the position of ephemeral little magazines in general: a position of (magnificent) failure.

Short-lived printed adventures – often starting with great ambitions and high energy, then quickly disappearing – offer an intriguing perspective on the cultural role of little magazines. Emerging in early twentieth-century experimental modernist print culture, little magazines have played a pivotal role in the proliferation of cultural and political groups and movements throughout the last decade, while also being associated with a marginal cultural position. As Suzanne Churchill and Adam McKible note, little magazines may be defined as "non-commercial enterprises founded by individuals or small groups intent upon publishing the experimental works or radical opinions of untried, unpopular, or under-represented writers."[2] Short-lived magazines in the ultimately marginal medium of little magazines thus occupy the position of being on the margins of the margins, in terms of both cultural position and status in the field of periodical studies.[3] I suggest, however, that short-lived and even one-time little magazines not only have a privileged role in the cultural ecology as temporary platforms for emerging writers and artists, but also may allow a better understanding of the medium of the little magazine as such.

The role of ephemeral magazines is particularly significant in the case of peripheral cultures, languages, or social positions, often subject to various publishing constraints – such as smaller readership markets and limited resources – which

Figure 3.1 *02010*, magazine cover and item (2010).

Source: Private collection.

make the production of long-lasting, sustainable platforms more difficult. While until recently periodical and little magazine scholarship focused almost exclusively on influential Anglo-American and trans-European magazines, the recent theoretical shift toward the study of "Modernisms" in the plural[4] sheds light not only on the diversity of little-platforms' international production, but also on the unique contribution of ephemeral, supposedly "failed" platforms to the history and theory of little magazines.

In this chapter I suggest that the disadvantages of marginal magazine makers[5] – either in terms of location, language, or social position – are often exploited when they choose to emphasize the merits of marginality and ephemerality, and even foster the notion of their own "failure." Examining little magazine makers' practices from this rhetorical-performative perspective enables observing failure as a desired ethos they commonly promote, rather than an unfortunate outcome to avoid (as expressed in the emblematic title "Practice More Failure" in the third issue of the art magazine *LTTR* in 2004).

To demonstrate the unique role of such ephemeral platforms and the notion of desired failures, specifically in a context of a complex, marginal cultural identity, I will address the field of Hebrew little magazines.[6] Dozens of short-lived Hebrew little platforms were established throughout the last two centuries, first in Europe by local writers and artists (mostly in Poland, Germany, and Russia but also outside of Europe in the US), later, since the early 20th century, also in Palestine, by immigrants who absorbed the thriving European little magazine culture and attempted, in many cases, to continue it in their new residence. This desire to maintain the European little magazine-fever in Palestine, nevertheless, has always been mixed with an attempt to create a local culture, corresponding with the new – utterly different – context and landscape, as well as with the immense difficulties raised by an unstable social and political environment. Thus, these attempts are to be read, as Chana Kronfeld suggests, in "the triple context of the internal Hebrew/Yiddish literary system, the Middle Eastern versions of cultural modernism, and the international modernist affiliations."[7] This inherent tension of European–Middle-Eastern magazine-making, is manifested in one case on which I will focus: a Hebrew magazine from 1964 entitled *Kiltartan* – a self-proclaimed "one-time periodical."

The oxymoronic notion of "one-time" or "nonperiodical" periodicals will eventually lead me to a conceptual differentiation between the terms "periodical" (governed by linear time) and "magazine" (governed by space, as I will demonstrate), and to advocate for a "magazine" rather than a "periodical" – that is, a spatial-synchronic rather than linear-diachronic – conception and study of small independent publishing platforms.

"Practice more failure": failure as ethos in the cultural ecology

Despite significant differences between artsy high-modernism booklets, independent literary periodicals, and rough stencil fanzines, they all share the ambition to present an independent alternative to institutionalized magazines and newspapers, to create small discourse communities, and most of all, to provide a platform of public expression for individuals or groups that would have difficulty finding one elsewhere. Nevertheless, the initiative as imagined is rarely realized, for an array of reasons, ranging from a simple lack of resources, to circumstantial life changes in the usually young makers, from frequent splits within creative groups, to limiting cultural factors such as censorship. Highly ambitious manifestos, replete with the usual proclamations, such as declarations on establishing a group or a movement and promising future actions, are often resolved in limited, short-lived adventures, leaving behind the manifestary speech acts as "unhappy," in J. L. Austin's terms;[8] that is, not meeting the actual conditions which enable them to work.

Failure, in its simplest, dictionary definition, is a relative, circumstantial term designating a gap between expectation and performance: "A failing to occur, be performed, or be produced; an omitting to perform something due or required."[9] A common occurrence in independent publishing, this gap makes the history of independent magazines to a large extent a history of (often impressive) failure. Indeed, the ultimate "success" or "failure" is often difficult to assess due to the vague expectations that makers evoke when launching a new platform, not always articulated in the form of decisive manifestary declarations. However, the notion of a little magazine as a "failure" is also derived from a second perspective: a normative, evaluative judgment, not necessarily assuming an unfulfilled expectation. Modern Western culture essentially associates "little" and "ephemeral" with "failure," while associating "wide-scale" and "long-lasting" with "success" – hence, the semantic connection between "successive" and "successful." Thus, short-lived, marginal little magazines occupy an inherently weak cultural position – a position of "failure" – even if they are not subject to any circumstantial gap between expectation and performance.

Although we indeed tend to perceive failure as a negative outcome that exposes weakness and inadequacy, in fact failure is also associated culturally with positive attributes, such as sincerity, ideological devotion, and even heroism (such as in the extreme case of death in battle). Failure is often conceived as an index of authenticity, and it potentially provides the "failed" endeavor with symbolic capital that could not have been created by the notion of success. For little magazine makers, the ethics of failure – which provide a seal of authenticity – can be much more rewarding than normative success (manifested in long-term existence, financial profit, or wide public acknowledgment). By marking themselves as failures, as I will demonstrate later, makers create an alternative authority: they appear subversive and noninstitutional, and separate themselves from governing power structures, market forces, and normative practices of production and distribution in the cultural field. This position allows them even to appear as cultural martyrs who sacrifice themselves "for a greater cause." In this sense, oxymoronic as it may seem, early self-marketing as "failures" enables little magazine makers to create a distinctive position, which is vital, as Pierre Bourdieu has shown, for one's positioning as an innovator in the cultural field.[10] Thus, ephemeral, marginal platforms carry a double set of potential cultural connotations: the negative hegemonic perception associating ephemerality with failure, and the potentially positive, or at least constructive and beneficial connotations of ephemeral "failures."

The constructive perception of supposed failures may be further strengthened from the scholarly perspective by observing the cumulative rather than the individual role of marginal short-lived magazines. The history of magazines is replete with ephemeral platforms, lasting only a few issues, often no more than a year, while nevertheless being highly influential as a whole. In fact, short-lived magazines comprised "the public face of modernism"[11] when the little magazine culture was just coming into being at the beginning of the twentieth century and had a significant role in creating cultural networks. Magazines such as *Simbolul* (1912), *Futuristy* (1914), *Blast* (1914–1915), and *Der Dada* (1919–1920), are only a few prominent examples out of dozens of modernist and avant-garde magazines, which had crucial functions for their contemporaries, as well as a substantial historical role as a whole, despite their short life span.[12] In "What Is Minor Poetry?" from 1943, T. S. Eliot highlighted the unique cultural significance of such short-lived ventures: "These small magazines often appear to circulate only among contributors and would-be contributors; their condition is usually precarious, they appear at irregular intervals, and their existence is brief. [. . .] Such groups," he writes, "frequently bind themselves together by formulating a set of principles or rules, to which usually nobody adheres; in course of time the group disintegrates." Nevertheless, he suggests, "their collective importance is out of all proportion to the obscurity in which they struggle."[13]

Ephemeral little magazines indeed bear such "collective importance," despite their arguably negligible individual value. They assemble a multiplicity of cultural forces, to a large extent independent of normative social barriers (though it must be said that the ability to produce even the simplest magazine assumes a certain degree of socioeconomic competence and a suitable habitus); thus they may be seen as valuable sources for gauging a cultural climate. These short-lived platforms are often heterogeneous, dynamic, and diverse, more immediate and attuned to the rhythm of daily life, often rapidly published and more responsive to changing sociopolitical issues than books, while also being less bound to a disciplined procedure of editing and self-censorship than successive publications such as newspapers and institutionalized periodicals. Supposedly "failed" magazines – either in terms of the gap between expectation and realization, or from the normative conception of "little" magazines as marginal, negligible products in the capitalist market of cultural goods – are in fact precious evidence of the cultural imagination of small communities at specific points in time. The documentation of these magazines, which often give voice to new artists and writers, and provide them with initially safe (albeit temporary) ground for expression, also potentially captures the point of departure of significant creative careers, as I will demonstrate here. As Martin Puchner suggests, the "success and failure" of independent cultural endeavors should not be measured by the realization of goals, but "by the force, inventiveness, and wit of these acts."[14] However, even if such qualities are doubtful, the "collective importance" of those initiatives remains.

Hence, ephemeral little magazines have a distinguished role in creating a diverse cultural ecology.[15] The ecological perception of little magazines in the cultural field may be complicit with the "conversation" and "party" models suggested by Churchill and McKible.[16] Both their interactive, anticombative models and the ecological model differentiate themselves from the common battle-line hypothesis, which assumes that little magazines are nourished mainly from artistic, literary, and political conflicts and are thus best described through the semantic field of competition and war. Rather, these alternative models emphasize little magazines' function in a complex network of connections and collaborations, while differentiating the commonly vociferous and combative rhetoric of little magazine makers from the dynamic and nonbinary ways the cultural field actually works. The ecological perception of culture highlights not only this notion of magazines as networks of interaction, but also the immense differences and variations of publishing platforms as forms of life – ranging from strong survivors (long-lasting platforms) to "ephemera" (originally derived from ecological jargon, as a term for "an insect that [in its imago or winged form] lives only for a day").[17] It is no coincidence that in a 1918 anonymous citation in *The Dial*, with which Churchill and McKible open their essay, magazines are described as "mushrooms or hardier plants [. . .] springing up overnight"[18] – that is, living creatures ranging from rapid ephemeral existence to a more rooted hold. The ecological perspective in this sense is particularly significant for understanding the role of marginal, ephemeral, and supposedly failed platforms. In cultural ecology, the notion of success is not valid for any individual creature as a separate entity, regardless of its life span, but only for the dynamic, synergetic, and prosperous function of the whole, and to the contribution of each and every creature to this synergetic texture.

Failure, from this perspective, cannot be an objective position in the cultural field, but it can certainly be a possible strategy for living and survival. Just as some actors in the cultural ecology present themselves in advance as significant and successful, as a positioning method, little magazine makers often embrace the opposite rhetoric towards a similar aim. Usually aware of their unique position as cultural entrepreneurs, as well as to the lurking danger of ephemerality and their normative marginal framing, magazine makers often embrace "failure" in advance as a performative act, a rhetorical gesture, which is, I suggest, the core of what might be referred to as a little magazine ethos.

Although seemingly self-contradictory indeed, the performance of failure is in fact an effective solution for the weak position of little magazines: if failure is regarded as a gap between expectation and performance, the rhetoric of failure-out-of-choice prevents such a gap from occurring, and paradoxically marks one as unable to fail. If failure is regarded as a normative judgment, the very choice to create a "little" magazine – an independent platform, limited in its scope, readership, and influence – is a demonstrative choice of what is normatively framed as a failure, as well as a way to question and challenge this conception. Even if most makers do not use the term "little magazine" explicitly, many essentially adopt the position of being "little" by emphasizing their marginality.

Such rhetoric of failure does not necessarily imply an actually demeaning self-perception of its makers. On the contrary, the defiant rhetoric of failure is perfectly compatible with the ethos of the avant-garde, shared with many little magazine makers, as an act of pioneering that is inevitably marginal and temporary – the avant-garde, after all, cannot preserve its sense of precedence (its "avant") over time. In fact, the only way to escape the ontological paradoxes of the avant-garde stance, as Dadaists well understood already in the second decade of the twentieth century, is the continual performance of failure. Thus, in Tristan Tzara's self-contradictory and self-annihilating manifestos, such as *Dada Manifesto 1918*, against the typical manifestary requirement by which "you have to want: ABC," he declares: "I am writing this manifesto and I don't want anything [. . .] Dada does not mean anything."[19]

Indeed, instead of offering a typical manifesto – a declaration of principles – ephemeral little magazines often present what I wish to define as a "declaration of failure": the activation of an array of textual and visual signals that imply the "failure" stance. Magazine makers foster the ethics of failure via various rhetorical techniques of strategic self-marginalization, which often include, as I will show, a rhetoric of self-reduction; self-sabotaging by way of self-contradictions; deliberate noneloquence, awkwardness, or oddness; and even overt declarations about a current or expected failure. "Failure," as a rhetorical action of ethos construction, plays a similar role even if the magazine eventually turns out to be "successful" in normative terms – that is, long-lasting and widely consumed.

Such rhetoric of failure appears in historical as well as contemporary magazines in various cultural contexts, however, as I wish to emphasize, it is particularly common when magazine makers occupy a marginal cultural identity, and thus must recruit unique rhetorical means to deal with their increased potential framing as failures. Before demonstrating extensively how such rhetoric of failure is activated in a historical magazine of a peripheral culture, in which the identity of the makers is complex, I will give two brief examples of this rhetoric in more contemporary magazines, and specifically, I wish to emphasize its role in the activity of women editors. Although this issue extends beyond the scope of this chapter, it must be mentioned that women cultural entrepreneurs are particularly subject to normative marginalization; thus it is not surprising that among the various unique rhetorical strategies they utilize[20] they also commonly employ the rhetoric of failure in their editorial language. However, instead of seeing the ethos of failure as a simple internalization of the hegemonic critical voice, I see it as a powerful performative tool that women writers use to increase the reliability and effectiveness of their action. In other words, women editors use the constructive potential of failure sophistically to position themselves effectively in the cultural ecology.

Such rhetoric of failure is found, for example, in the Hebrew magazine *Rooms* (*Hadarim*, edited by Helit Yeshurun in Tel Aviv), a surprisingly long-lasting platform in retrospect, in which the editor declared in the opening antimanifesto of the first issue from 1981: "Publishing a new periodical of poetry now, I'm aware of its absurdity, the over-indulgence of it. Who needs it? I can't even point at a void, an open door, a void that needs to be filled. It is not missing."[21] In this case, as in many others, the editor presents an apologetic opening that seemingly faults the *raison d'etre* of the new platform and presents it as a predetermined failure, but in fact marks it as a modest, altruistic quest, and creates a sharp contrast to the pretentious rhetoric of mainstream periodicals.

Similarly, in her New York–based zine, *I Dreamed I Was Assertive* (second issue, 1999), as if to diminish and undervalue her own platform, Celia Perez writes: "I hate the title of this Zine, so this will probably be the last issue of it in its 'I Dreamed I Was Assertive' incarnation. Not that anyone cares."[22] The writer situates her zine as a predetermined failure – marginal, negligible, and overall worthless – but at the same time fixes her creative voice as authentic, sincere, and reliable.

This type of "declaration of failure" through self-deprecation is only one strategy in the repertoire. In other cases, even if little magazines seem to celebrate their launch flamboyantly, they often make significant efforts to perform failure via various rhetorical devices, as is the case with *Kiltartan*.

Little ephemeral creatures: the case of *Kiltartan*

To illustrate the notion of the performance of failure, and the cultural importance of so-called failures, particularly in a complex cultural context, I refer to a rare and unusual Hebrew example that nevertheless encapsulates the issues at play in many ephemeral, marginal magazines: the magazine *Kiltartan*.[23]

This magazine, published in Tel Aviv in 1964, was edited by three partners: the poet Maxim Gilan – a radical leftwing activist born in Lille, France, who escaped to Palestine in 1944 after spending years in Fascist Spain; artist and architect Nachum Cohen (named Kane in the magazine) – born in Sofia, Bulgaria, who immigrated to Israel in its year of establishment (1948); and Moscow-born poet Meir Wieseltier – who also immigrated in the late 1940s, and coedited the magazine until a point of division occurred between the editors, at which point he decided to quit. *Kiltartan* was named after the beloved getaway of W. B. Yeats in Ireland, which is alluded to in his poem "An Irish Airman Foresees His Death" from 1919, a poem centered around the theme of ephemerality, written from the perspective of a soldier before his predictable death in battle:

> I know that I shall meet my fate
> Somewhere among the clouds above;
> Those that I fight I do not hate,
> Those that I guard I do not love;
>
> My country is Kiltartan Cross,
> My countrymen Kiltartan's poor,
> No likely end could bring them loss
> Or leave them happier than before.
>
> Nor law, nor duty bade me fight,
> Nor public men, nor cheering crowds,
> A lonely impulse of delight
> Drove to this tumult in the clouds;
>
> I balanced all, brought all to mind,
> The years to come seemed waste of breath,
> A waste of breath the years behind
> In balance with this life, this death.[24]

This demonstrative choice of a foreign and cosmopolitan-sounding title for the magazine, resonating with international modernism but unfamiliar to most of the local readers, reflects the unique position of *Kiltartan*'s makers: young, educated urban immigrants in their twenties, not only familiar with the European avant-garde little magazine publishing culture, but also aspiring to recreate it in the Tel Aviv of the early 1960s. This endeavor encompassed inherent paradoxes: a belated engagement with the historical avant-garde, as well as an attempt to "import" European creative patterns to a Middle Eastern cultural and linguistic landscape, thereby creating an impossibly anomalous motivational mixture, but at the same time triggering an extremely creative and original venture.

Aware of the paradoxical nature of their initiative, and inspired by Yeats's poem, the editors chose to define the magazine from the outset as a temporary experiment, an oxymoronic "one-time" periodical (the Hebrew word for "periodical" is "ktav-et": a hybrid of "writing" and "time"), published as a special edition of 500 copies. Only a few of these copies, circulating among random collectors and auction houses, remain, and to date *Kiltartan* is not included in any official collection or archive, with only one copy in the Israeli National Library. Notwithstanding its nearly invisible status (which reflects the state of Hebrew little magazines in general: an uncharted field to a large extent),[25] *Kiltartan* presents a unique example of an aesthetically experimental magazine that links itself to a historical tradition of making little magazines, while acting out an ethos of failure.

The performance of failure is already manifested in *Kiltartan*'s cover: a plain brown cardboard folder ironically decorated with black ribbon and a red wax seal (Figure 3.2). The seal, conveying a sense of a praised accomplishment, clashes with the ascetic brown cover, and hence functions as a costume: a mask of exaggerated and ironic self-importance. The explicit declaration of ephemerality (being a "one-time" thing) right on the first page, was followed by the teasing declaration: "The editorial staff is not responsible for its actions." (Figure 3.3). From then on, the rhetoric of *Kiltartan*, in its eclectic texts and images, its rough craft style (the items are printed on eclectic stenciled papers in various sizes and colors, replete with handwritten additions), and its overall tone, celebrates its marginality, and in fact highlights what is framed as a moment of failure. "*Kiltartan*," the editors declare in one of their manifestary texts, "is the only periodical that is not trying to set a precedent."

Kiltartan – doubtlessly influenced by Dada magazines such as *Merz* (edited by Kurt Schwitters in Hannover and appearing irregularly between 1923 and 1932), as well as by the long-lasting postwar American *MAD Magazine* – was comprised of an eclectic, interdisciplinary, carnivalesque collection of materials: works of literature and art, graphic experiments, manifestos, and collaborations with artists.[26] Some items were more explicitly political (such as a provocative call to grant Ezra Pound the Nobel Prize despite being a Fascist), others protesting against the bourgeois literary field. The holistic, bemusing, and defiant stance of the magazine, and its ethics of failure, may be epitomized by three of the items, all unusual components of a magazine at the time in terms of genre: a comic strip, a lottery card, and a map.

The comic strip presented in *Kiltartan* (Figure 3.4) consisted of a pastiche of a French sailor comics. While it presented images of sailors occupied with hard

Figure 3.2 *Kiltartan*, magazine cover (1964).
Source: Private collection.

labor, in the text bubbles the figures cite lines from Hebrew poetry – creating an ironic gap between the solemn faces and the lyrical phrases they utter, as well as between the European-style drawings and the ultimately local, internal literary discourse. The comics ends with the conventional declaration "to be continued," while, as mentioned, the magazine was defined from the outset as a one-time publication. Thus, the item created an amusing tension between the requirement for continuity and the ethos of ephemerality, deliberately, and demonstratively creating an unfulfilled ("failed") promise.

Another item, a lottery card entitled "How to Build Culture," asked the reader to gamble and write which poets will receive literary prizes in the following decade. It provided a list of popular and appreciated poets, as well as some fictional names, and asked the reader to "consider the following criteria: 1. Details known to you about the poet; 2. Average annual production; 3. The level of lyricism; 4. Standard deviation; 5. Other customary criteria." This item clearly mocks the bourgeois

Figure 3.3 Kiltartan, opening page (1964).

Source: Private collection.

Figure 3.4 Kiltartan, comics (1964).
Source: Private collection.

institutionalization of literature, and positions *Kiltartan* as an authentic alternative. *Kiltartan* seemingly declares it is "not playing the game" with its refusal to pursue traditional rites of reception, which it sees purely as games of reputation that also inevitably require commitment to the long-term establishment – refusing to go this route even at the cost of being figured as a failure (i.e., not deserving of any literary prize). This participatory item also reflects an attempt to break the hierarchy between maker and reader in a playful, clownish tone, as if criticizing the magazine's monolithic authoritative voice.

This overall humoristic attitude nevertheless enabled the makers to conceal a relatively radical leftwing approach, questioning the Zionist ethos, mocking and parodying Israel's national ethos and symbols. An item presenting a satirical map of the biblical land of Israel, composed of fragments of the land framed in various historical and pseudo-historical imagery, was unusually daring. In the early 1960s, Israel was still a young state preoccupied with the mission of nation-building, and popular discourse, even in bohemian circles, was saturated with military jargon and for the most part not explicitly critical of the national politics. In fact, *Kiltartan* was one of the few cultural products to voice this kind of "heresy" at the time, and its declared temporary and marginal status was one of the features that enabled such a stance.[27]

But perhaps the most interesting aspect of *Kiltartan* as a temporary creative space is its participants. Although the magazine has been described before in traditional combative terminology as a platform for literary battles and control over the cultural field, from the ecological perspective it mostly served as an ephemeral, and a highly productive ground for interaction and collaboration.[28] *Kiltartan* hosted three Hebrew poets who would become some of the most appreciated voices of the later 1960s: Yona Wallach, Meir Wieseltier, and Yair Hurvitz. It was one of the first publishing platforms for these young writers, who were later among the leading cultural figures of their time. Yona Wallach, among the most original and influential Hebrew woman poets of all time, published in *Kiltartan* one of her first poems – an untitled poem later published under the title "Kasius" – when she was only nineteen.

The three poets, who would become life-long friends, first met in 1963, in an experimental evening of performative readings that heralded the appearance of *Kiltartan* before the magazine even existed, blurring the boundaries between literature and performance, and challenging the conservative poetry reading events of the period, which typically consisted of a reading on stage in front of a seated audience. Influenced by the spirit of early avant-garde performances such as the mythological Dada shows at Cabaret Voltaire in Zurich in 1916, this evening included a theatrical performance with costumes, reading poetry while smoking and drinking on stage, projections of collages of Tel-Aviv landscapes, and Constructivist music.[29] The original plan (fostered mostly by Maxim Gilan, the French-born cultural entrepreneur and political activist who coedited the magazine), was creating what they termed an "action-house" called *Kiltartan*. Thus, the magazine was presented as a cultural space containing multiple potential activities (Figure 3.5).

Kiltartan, a multidisciplinary "action-house" as defined by its makers, was in many ways ahead of its time – a unique example of an avant-garde platform in Hebrew in the early 1960s – and at the same time it was perfectly contemporary, appearing simultaneously with international neo-avant-garde movements such as the Situationists and Fluxus – both active in the early 1960s. Without being in direct contact with these movements, and probably not even fully aware of their parallel work, the *Kiltartan* makers demonstrated surprising similarities

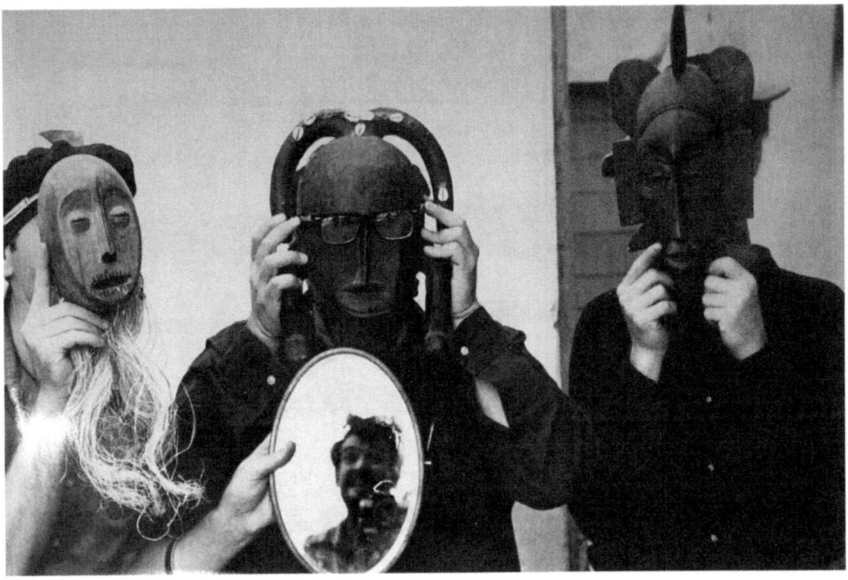

Figure 3.5 *Kiltartan*, participants (1963–1964).
Source: Amnon Weinstein.

in their output to what was being created by these other movements. *Kiltaran*'s "action-house" activity resonated with situationist "happenings" and "situations" (fostered by French thinker and artist Guy Debor), and a manifestary letter that appeared in *Kiltartan* resembled aesthetically, in its alternating pattern of black and white, as well as in the theme of creative dynamism, the *Fluxus* manifesto by George Maciunas, written in the very same year (Figure 3.6). Most of all, the desire to escape institutionalized procedures through ephemeral, dynamic actions created a direct link between these neo-avant-garde groups.

Kiltartan as "action-house" indeed played an important role as a temporary shelter for emerging writers, but nevertheless remained marginal and to a large extent forgotten through time, often not even mentioned as a milestone of its participants' careers. Despite being excluded from the literary canon, the rumor of *Kiltartan* circulated among young writers and artists, who could not always even find an actual copy. The early 2000s (particularly starting in 2005 with the parallel publication of several new prominent magazines) marked a revival in the field of Hebrew little magazines, which triggered a look back in search of historical publishing models and provided fertile ground for a renewed interest in *Kiltartan*. Young Hebrew writers at the beginning of the 2000s, thirsty for models of the avant-garde in Hebrew and motivated to create little magazines of their own, were inspired by this rare magazine of the early 1960s, using it to create a micro-genealogy of Hebrew avant-garde magazine makers.

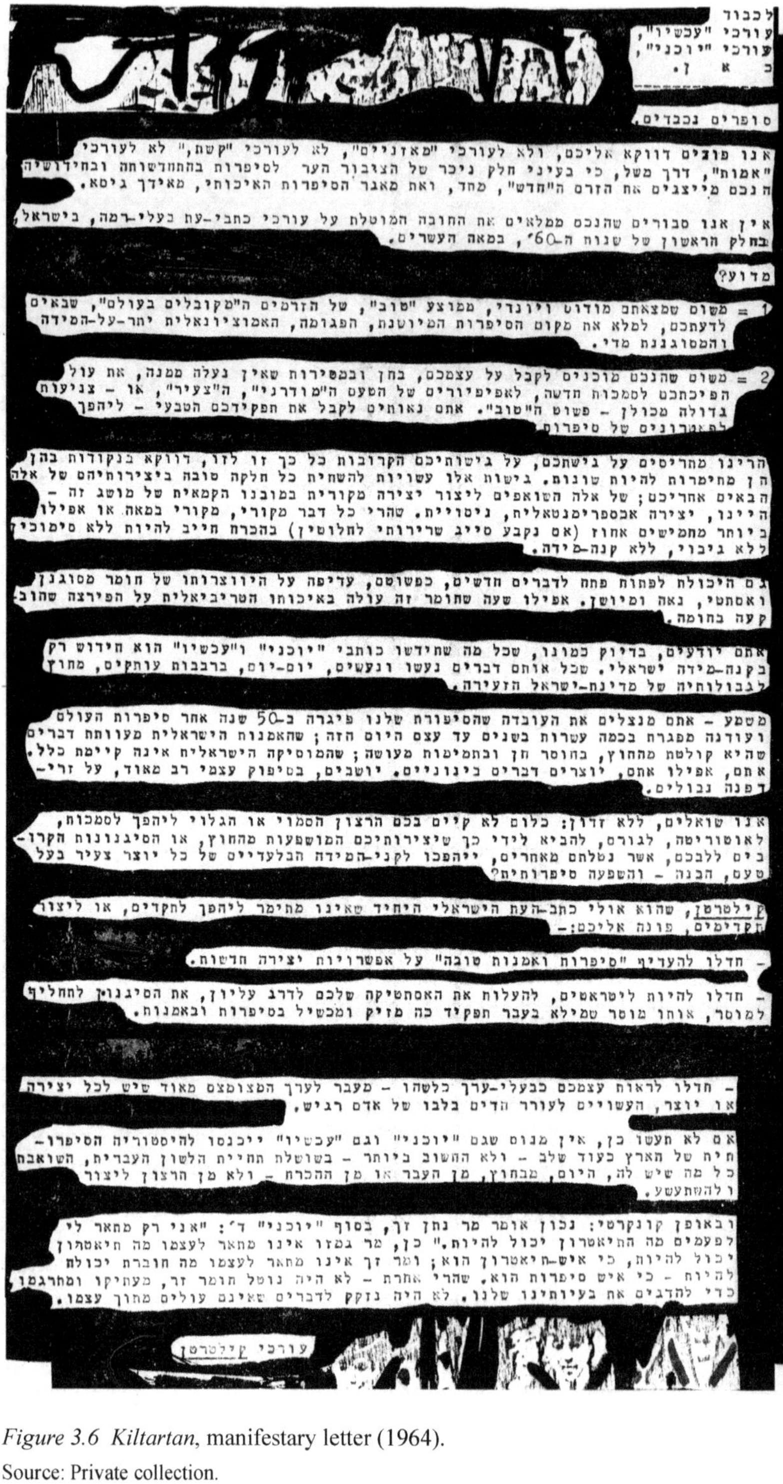

לכבוד
עורכי "עכשיו",
עורכי "יוכני",
כ א ן.

סופרים נכבדים,

אנו פונים דווקא אליכם, ולא לעורכי "מאזניים", לא לעורכי "קשת," לא לעורכי "אמות", דרך משל, כי בעיני חלק ניכר של הצבור הער לסיפרות בהתחדשותה ובחידושיה הנכם מייצגים את הזרם ה"חדש", מחד, ואת מאגר הסיפרות האיכותי, מאידך גיסא.

אין אנו סבורים שהנכם ממלאים את החובה המוטלת על עורכי כתבי-עת בעלי-רמה, בישראל, בחלק הראשון של שנות ה-60', במאה העשרים.

מדוע?

1 = משום שמצאתם מודוס ויונדי, ממוצע "טוב", של הזרמים ה"מקובלים בעולם", שבאים לדעתכם, למלא את מקום הסיפרות המיושנת, הפגומה, האמוציונאלית יתר-על-המידה והמסוגננת מדי.

2 = משום שהנכם מוכנים לקבל על עצמכם, בחן ובמסירות שאין נעלה ממנה, את עול הפיכתכם לסמכות חדשה, לאפיפיורים של הטעם ה"מודרני", ה"צעיר", או - צניעות גדולה מכולן - פשוט ה"טוב". אתם נאותים לקבל את תפקידכם הטבעי - ליהפך לפאטרונים של סיפרות

הרינו מהריסים על גישתכם, על גישותיכם הקרובות כל כך זו לזו, דווקא בנקודות בהן הן מתימרות להיות שונות. גישות אלו עשויות להשחית כל חלקה טובה ביצירותיהם של אלה הבאים אחריכם; של אלה השואפים ליצור יצירה מקורית במובנו הקמאית של מושג זה - היינו, יצירה אכספרימנטאלית, ניסויית. שהרי כל דבר מקורי, מקורי במאה או אפילו ביותר מחמישים אחוז (אם נקבע סייג שרירותי לחלוטין) בהכרח חייב להיות ללא סימוכין ללא גיבוי, ללא קנה-מידה.

גם היכולת לפתות פתח לדברים חדשים, כפשוטם, עדיפה על היווצרותו של חומר מסוגנן ואסתטי, נאה ומיושן. אפילו שעה שחומר זה עולה באיכותו הטריביאלית על הפירצה שהובקעה בחומה.

אתם יודעים, בדיוק כמונו, שכל מה שחידשו כותבי "יוכני" ו"עכשיו" הוא חידוש רק בקנה-מידה ישראלי. שכל אותם דברים נעשו ונעשים, יום-יום, ברבבות עותקים, מחוץ לגבולותיה של מדינת-ישראל הזעירה.

משמע - אתם מנצלים את העובדה שהסיפורת שלנו פיגרה ב-50 שנה אחר סיפרות העולם ועודנה מפגרת בכמה עשרות בשנים עד עצם היום הזה; שהאמנות הישראלית מעוותת דברים שהיא קולטת מהחוץ, בחוסר חן ובתמימות מעושה; שהמוסיקה הישראלית אינה קיימת כלל. אתם, אפילו אתם, יוצרים דברים בינוניים. יושבים, בסיפוק עצמי רב מאוד, על זרי-דפנה נבולים.

אנו שואלים, ללא זדון: כלום לא קיים בכם הרצון הסמוי או הגלוי ליהפך לסמכות, לאוטוריטה, לגורם, להביא לידי כך שיצירותיכם המושפעות מהחוץ, או הסיגנונות הקרובים ללבכם, אשר נטלתם מאחרים, ייהפכו לקני-המידה הבלעדיים של כל יוצר צעיר בעל טעם, הבנה - והשפעה סיפרותית?

קילטרטן, שהוא אולי כתב-העת הישראלי היחיד שאינו מתימר ליהפך לתקדים, או ליצור תקדימים, פונה אליכם:-

- חדלו להעדיף "סיפרות ואמנות טובה" על אפשרויות יצירה חדשות.

- חדלו להיות ליטראטים, להעלות את האסתטיקה שלכם לדרג עליון, את הסיגנון להחליף למוסר, אותו מוסר שמילא בעבר תפקיד כה מזיק ומכשיל בסיפרות ובאמנות.

- חדלו לראות עצמכם כבעלי-ערך כלשהו - מעבר לערך המצומצם מאוד שיש לכל יצירה או יוצר, העשויים לעורר הדים בלבו של אדם רגיש.

אם לא תעשו כן, אין מנוס שגם "יוכני" וגם "עכשיו" ייכנסו להיסטוריה הסיפרותית של הארץ כעוד שלב - ולא החשוב ביותר - בשושלת תחיית הלשון העברית, השואבת כל מה שיש לה, היום, מבחוץ, מן העבר או מן ההכרח - ולא מן הרצון ליצור ולהשתעשע.

ובאופן קונקרטי: נכון אומר מר נתן זך, בסוף "יוכני" ד': "אני רק מתאר לי לפעמים מה התיאטרון יכול להיות." כן, מר גמזו אינו מתאר לעצמו מה תיאטרון יכול להיות, כי איש-תיאטרון הוא; ומר זך אינו מתאר לעצמו מה חוברת יכולה להיות - כי איש סיפרות הוא. שהרי אחרת - לא היה נוטל חומר זר, מעתיקו ומתרגמו כדי להדביק את בעיותינו שלנו. לא היה נזקק לדברים שאינם עולים מתוך עצמו.

עורכי קילטרטן

Figure 3.6 Kiltartan, manifestary letter (1964).

Source: Private collection.

One of these *Kiltartan*-inspired magazines was the one-time magazine with which I began this discussion: *02010* (the title stands for the publication year, 2010, while also implying a subversion of the normative way in which years are counted, toward a sense of trans-humanistic time).[30] This magazine, similarly to *Kiltartan*, presented a cardboard folder comprising eclectic literary works and manifestos, as well as the mentioned group photo of the Columbia crew. Beyond the ironic, campy facade, the choice to present this photo can now be seen as the perfect image of a performance of failure: the makers celebrate Columbia's moment of collective glory before its colossal defeat, while also creating a somewhat provocative parallel to the tragic irony of the photo with their own existence as cultural entrepreneurs. "We are aiming for the moon," they seem to say, but at the same time we are aware of our inevitable ephemerality and deliberately choose it, somewhat like the airman who foresees his death in Yeats's "Kiltartan" poem. With this photo, this highly marginal magazine celebrated its own ephemeral moment – a moment of magnificent failure – while also implying a transnational community of short-lived entrepreneurs. This also implies that for its publishers, a magazine is often as precious as a living creature, and symbolically, the act of publishing is a matter of life and death.

Highlighting this existential perception of print, the legendary short-lived periodical *The Awakener* (Ha'meorer, London 1906), edited by the most praised Hebrew writer of the twentieth century, Yosef Haim Brenner – a well-known master of the performance of failure, opened with this statement:

> See, the life and death of this platform are in your hands. At your will – it will spread and flourish in quality and quantity, and at your will – it will wilt and wither and fade away. You are the only master! [. . .] "What is this new creature that you are about to create? What is its name?" – We don't want to name something that hasn't even begun its youth. [. . .] And all these dreams, all these lives are in your hand, reader. Please, be careful with life coming into being.[31]

This request from the reader – "be careful with life coming into being" – at the moment of launching a new magazine when its future is still unknown, highlights the fragile moment of the early steps of a publishing platform and situates failure as a constantly lurking danger. It is exactly this "failure anxiety" that magazines such as *Kiltartan* and *02010* transform into a performative act of failure-by-choice.

The performance of failure may be deceptive: it creates the impression that the makers resist being documented and perpetuated. The ethos of failure allegedly reflects their desire to be "out of time," and accordingly, outside the archive, as if performing Derrida's oxymoronic death-driven "Archive Fever": for Derrida the archival act marks an intersection of conflicting desires and anxieties, oscillating between preservation, commemoration, and a deadly fixation.[32] Scholars and archivists, for their part, should beware of duplicating the failure rhetoric of makers, and rather persistently pursue little magazines' documentation and preservation. Brenner's request to handle printed matter carefully when it has just been brought into being, can be transformed to a similar request concerning magazines' afterlife: if little magazines are evidence of small slices of life, their documentation, archiving, and digitization is vital for the preservation of this cultural ecology.

Beyond failure: "periodical" vs. "magazine" and the *mise-en-abyme* in the archive

The discussion about the role of marginal, ephemeral platforms can benefit from a second glimpse at the terms "periodical" versus "magazine." These terms are often used as synonyms, and although there is indeed no rigid taxonomic differentiation between the two, they ultimately reflect two different ontological trajectories.

The term "periodical" obviously highlights a sense of time – a "period." A periodical is a publication that is supposed to appear at regular intervals, or at least "from time to time." When defining a printed product as a periodical, that is, in terms of linear time, ephemerality necessarily equals failure, as it creates unfortunate, oxymoronic "nonperiodical periodicals."

The term "magazine," on the other hand, highlights a sense of space in its Semitic etymological roots – a "magazine" is something that stores (echoing the French word for store: "magasin." Also: a weapon's storage compartment for ammunition). The word "magazine" is derived from the Arabic term (also used in Hebrew) "Mahzan" or "Mahsan" (مخزن) – meaning a storage place, a small warehouse. Metaphorically, a magazine is thus a cultural space. This concept is commonly reflected in magazine makers' rhetoric, when they describe the magazine as a "home," a "room," a "shelter," or an "address," such as in the case of *Kiltartan*, aspiring to function first and foremost as an "action-house." Many Hebrew magazines of the last few decades imply this metaphor in their title, such as *Hadarim* ("Rooms," 1981), *Makom* ("Place," 1985), *Gag* ("Roof," 1998), and *Ketovet* ("Address," 2009).

Makers indeed often struggle between a "periodical" and a "magazine" identity: between a desire for linear progress and survival through time, and an ambition to function as a temporary "place" – a shelter for a collection of artists, writers, and works. While they often fail in being "periodical," it seems that they simply cannot come short in serving as a "magazine:" providing a temporary space, an ephemeral storage or shelter.

A "magazine" rather than a "periodical" perception of printed matter may thus allow us to focus on the synchronic function of ephemeral publishing platforms, rather than on their diachronic, linear progress through time and their successive existence. A diachronic examination is obviously vital for evaluating the overall impact of a cultural product. However, this impact can be revealed only when paying attention to its function in a cultural ecology at a given point in time. Short-lived platforms often seem to have a very small impact "periodically," if at all, and thus may be framed as negligible contributions, as failures, yet their "magazine" impact – their function as a cultural space synchronically – provides the necessary groundwork for later cultural developments, as in the case of *Kiltartan* presented here.

There is also another possible implication of the spatial reading of magazines; examining a magazine as space may emphasize the unique, symbolic place of magazines in archives. If a "magazine" is defined by its function as a "Mahsan" – a storage place for various items, something that collects, summons, and presents – a "magazine" is in fact a little archive. Thus, when a magazine is placed *in* an archive, it does not only function as a unified object for preservation and demonstration, but also mirrors the very act of archiving. In fact, similarly to the spatial designation of "magazine," the "archive," as Derrida notes, is ultimately a space, an "arkheion": "a house, a domicile, an address."[33] The magazine is thus a *mise-en-abyme* inside the archive – a tiny model that captures the essence of the archive as such. It is thus, perhaps, the most representative object of the archive.

These implications of the spatial-temporal conceptualization, may serve as small methodological steps towards a better understanding of the function of publishing platforms, which go beyond traditional dichotomies of "marginal" vs. "influential," or "failed" vs. "successful." The differentiation between little magazines as small cultural spaces (or "action-houses"), which create horizontal networks, as opposed to periodicals as linear enterprises best characterized by vertical processes and long-term progress (or "mushrooms" vs. "hardier plants"), may present a possible key to the sought-after taxonomy in the developing field of periodical/magazine studies.[34]

Notes

1 A version of this work was first presented in "Summoning the Archive: A Symposium on the Periodical, Printed Matter, and Digital Archiving," organized by Meghan Forbes at the Institute for Public Knowledge, NYU, on 11–13 of May, 2017. I thank Meghan Forbes for the dedicated and productive editing, and for her observant and illuminating insights.

2 Suzanne W. Churchill and Adam McKible, "Introduction," in *Little Magazines and Modernism: New Approaches* (Aldershot and Burlington: Ashgate, 2007), 6.

3 Starting in the late 1970s and increasingly in the last few decades, multiple studies have been dedicated to "periodical studies." See: Sean Latham and Robert Scholes, "The Rise of Periodical Studies," *PMLA* 121, no. 2 (2006): 517–531. Within this emerging field, there has been growing interest in the role of modernist little magazines. See, among others, the early account of Malcolm Bradbury and James McFarlane, "Movements, Magazines and Manifestos: The Succession from Naturalism," in *Modernism: 1890–1930*, ed. Malcolm Bradbury and James McFarlane (Middlesex: Penguin, 1976), 192–205; and the more recent work of Robert Scholes and Clifford Wulfman, eds., *Modernism in the Magazines: An Introduction* (New Haven: Yale University Press, 2010). In recent years, the critical attention also shifted to local print and little magazine cultures. See, among others: Eric B. White, *Transatlantic Avant-Gardes: Little Magazines and Localist Modernism* (Edinburgh: Edinburgh University Press, 2013).

4 This relatively new terminology which has become widespread during the last decade, manifests in titles such as: Peter Brooker, Andrzej Gasiorek, and Deborah Longworth, eds., *The Oxford Handbook of Modernisms* (New York: Oxford University Press, 2010).

5 I use the term "makers" (as opposed to "editors," for example), to emphasize the dimension of multidisciplinary craft involved in the making of many little magazines, particularly in the case of limited-edition magazines.

6 While Hebrew Modernism has been studied extensively, the field of Hebrew periodicals and little magazines is markedly understudied. Although a few scholars have dedicated parts of their work to the study of specific Hebrew periodicals, and some work has been dedicated to Hebrew fanzines, the field of Hebrew little magazines, particularly in the context of international magazine production, remains to a large extent uncharted. Most magazines have not yet been systematically archived or digitized. See Nurit Govrin's marked study of Hebrew periodicals, and specifically her comments about short-lived platforms: Govrin Nurit, *Reading the Generations* (Tel Aviv: Gvanim, 2002), 43–67.

7 Chana Kronfeld, *On the Margins of Modernism* (Berkeley: University of California Press, 1996), 65.

8 John Langshaw Austin, *How to Do Things with Words* (Oxford: Oxford University Press, 1962), 14.

9 "Failure, N," *OED Online*, March 2018. Oxford University Press, accessed April 18, 2018, www.oed.com/view/Entry/67663?redirectedFrom=failure.

10 Pierre Bourdieu, *The Field of Cultural Production* (Cambridge: Polity Press, 1993), 40–84, 116.

11 Mark S. Morrisson, *The Public Face of Modernism: Little Magazines, Audiences, and Reception, 1905–1920* (Madison: University of Wisconsin Press, 2001).

12 For lists of early avant-garde magazines, in which short-lived platforms occupy a significant place, see digital reservoirs such as Avant-garde and Modernist Magazines, https://monoskop.org/Avant-garde_and_modernist_magazines and The Modernist Journals Project, http://modjourn.org/.

13 Thomas Stearns Eliot, *On Poetry and Poets* (New York: Farrar, Straus and Giroux, 2009), 35–36.

14 Martin Puchner, "It's Not over ('Til It's over)," in "What Is an Avant-Garde?" *New Literary History* 41 (2010): 917.

15 See discussion on the ecology of literature as a theoretical model in: Avidov Lipsker, *Ecology of Literature* (Ramat Gan: Bar Ilan University Press, 2018).

16 Suzanne W. Churchill and Adam McKible, eds., "Introduction," in *Little Magazines and Modernism: New Approaches* (New York: Routledge, 2016), 12–14.

17 "Ephemera, N," *OED Online*, March 2018. Oxford University Press, accessed April 18, 2018, www.oed.com/view/Entry/63197?rskey=mbJszH&result=1&isAdvanced=false.

18 Churchill and McKible, "Introduction," 3.

19 Mary Ann Caws, *Manifesto: A Century of Isms* (Lincoln: University of Nebraska Press, 2001), 297, 300, 301.

20 See an extensive discussion on women editors' rhetoric (such as the use of an "ironic editorial 'we',") in: Jayne E. Marek, *Women Editing Modernism: "Little" Magazines and Literary History* (Kentucky: The University Press of Kentucky, 1995).

21 Hillit Yeshurun, ed., *Hadarim*, vol. 1, Tel Aviv, Galeria Gordon,1981. Unless otherwise noted, all translations from the Hebrew are my own – NA.

22 Celia Perez, ed., *I Dreamed I Was Assertive*, vol. 2, New York,1999, n. p.

23 See a first full account of this magazine in: Nana Ariel, "Kiltartan: A One-Time Periodical: On Ephemeral Magazines and Modernist Moves," *OT: A Journal of Literary Criticism and Theory* 6 (2016): 159–190. [Hebrew].

24 William Butler Yeats, "An Irish Airman Foresees His Death," *The Wild Swans at Coole* (New York: The Macmillan Company, 1919): 13.

25 See note 6.

26 Among the artist participants: Nitzan Aviv who contributed images and prints throughout the booklet, Amnon Weinstein, a violin maker who presented in *Kiltartan* an exposition of assemblages, and the well-known Cenaani artist Yitzhak Danziger, who is interviewed in the booklet.

27 The very few reviews *Kiltartan* received for the most part treated it as a marginal and bizarre phenomenon and did not perceive its political criticism as a threat. See Ariel, "Kiltartan," 181.
28 In a detailed essay on *Kiltartan* I have shown that the perception of *Kiltartan* through a conventional combative model tends to replicate the magazine's own combative rhetoric and at the same time misses its complex functions in the cultural ecology. See Ariel, "Kiltartan," 184–190.
29 Based on a series of interviews I conducted with Nachum Cohen, coeditor of *Kiltartan*, Tel Aviv, October 2014.
30 The futuristic connotations of *02010* were further emphasized in another magazine established by the same editor, Oded Carmeli, in 2011, entitled "Hava Lehaba" (in free translation: "come on to the future"), a meta-modernist trans-humanistic oriented magazine.
31 Yosef Haim Brenner, ed., *Hameorer*, vol. 1, London, Masada, 1906. (The text was written by Brenner's editing partner Yehoshua Radler Feldman.)
32 Jacque Derrida, "Archive Fever: A Freudian Impression," *Diacritics* 25, no. 2 (1995): 14–19.
33 Ibid., 9.
34 Churchill and McKible, "Introduction," 8.

Bibliography

Ariel, Nana. "Kiltartan: A One-time Periodical: On Ephemeral Magazines and Modernist Moves." *OT: A Journal of Literary Criticism and Theory* 6 (2016): 159–190.

Austin, John Langshaw. *How to Do Things with Words*. Oxford: Oxford University Press, 1962.

Bourdieu, Pierre. *The Field of Cultural Production*. Cambridge: Polity Press, 1993.

Bradbury, Malcolm, and James McFarlane. "Movements, Magazines and Manifestos: The Succession from Naturalism." In *Modernism: 1890–1930*, edited by Malcolm Bradbury and James McFarlane, 192–205. Middlesex: Penguin, 1976.

Brenner, Yosef Haim, ed. *Hameorer*, Vol. 1. Masada, London, 1906.

Brooker Peter, Andrzej Gasiorek, and Deborah Longworth, eds. *The Oxford Handbook of Modernisms*. New York: Oxford University Press, 2010.

Caws, Mary Ann. *Manifesto: A Century of Isms*. Lincoln: University of Nebraska Press, 2001.

Churchill, Suzanne W., and Adam McKible, eds. *Little Magazines and Modernism: New Approaches*. Aldershot and Burlington: Ashgate, 2007.

Derrida, Jacque. "Archive Fever: A Freudian Impression." *Diacritics* 25, no. 2 (1995): 9–63.

Eliot, Thomas Stearns. *On Poetry and Poets*. New York: Farrar, Straus and Giroux, 2009.

Gilan, Maxim, Nachum Cohen, Meir Wieseltir, et al. *Kiltartan*, Tel Aviv, Kiltartan Action House,1964.

Govrin, Nurit. *Reading the Generations*. Tel Aviv: Gvanim, 2002.

Kronfeld, Chana. *On the Margins of Modernism*. Berkeley: University of California Press, 1996.

Latham, Sean, and Robert Scholes. "The Rise of Periodical Studies." *PMLA* 121, no. 2 (2006): 517–531.

Lipsker, Avidov. *Ecology of Literature*. Ramat Gan: Bar Ilan University Press, 2018.

Marek, Jayne E. *Women Editing Modernism: 'Little' Magazines and Literary History*. Kentucky: The University Press of Kentucky, 1995.

Morrisson, Mark S. *The Public Face of Modernism: Little Magazines, Audiences, and Reception, 1905–1920*. Madison: University of Wisconsin Press, 2001.
Perez, Celia, ed. *I Dreamed I Was Assertive*, New York, Vol. 2, 1999.
Puchner, Martin. "It's Not Over ('til It's Over)." In "What Is an Avant-Garde?" *New Literary History* 41 (2010): 916–928.
Scholes, Robert, and Clifford Wulfman, eds. *Modernism in the Magazines: An Introduction*. New Haven: Yale University Press, 2010.
White, Eric B. *Transatlantic Avant-Gardes: Little Magazines and Localist Modernism*. Edinburgh: Edinburgh University Press, 2013.
Yeats, William Butler. "An Irish Airman Foresees His Death." *The Wild Swans at Coole*. New York: The Macmillan Company, 1919, 13.
Yeshurun, Hillit, ed. *Hadarim*, Tel Aviv, Galeria Gordon, Vol. 1, 1981.

Part II

Object

4 Via postal

Networked publications in and out of Latin America[1]

Zanna Gilbert

"Creativity is necessarily anti-social. This is mitigated by the idea of the magazine – communication,"[2] wrote the artist and organizer David Mayor in the very first issue of the 1970s magazine *Schmuck*, published by the Beau Geste Press in the county of Devon, in the United Kingdom. Mail art is often characterized as an intimate one-to-one communication, but Mayor's comment reveals that there was an impetus to create a public sphere in the mail art movement. This chapter explores how publications were one of the pivotal networking strategies that enabled mail artists to exchange artworks and information across borders, allowing for international communication and for the creation of a subcultural public sphere. The magazines, journals, zines, and mailers that circulated within the mail art network articulated ideas about communication, mass media, anti-institutionalism, the boundaries of artistic practice, and the central principles of publishing in and of themselves. The publications Hexágono '71 (Edgardo Antonio Vigo, La Plata, Argentina); Commonpress (founded by Pawel Petasz in Poland, subsequently revolving editors and nations); Schmuck (Beau Geste Press, Devon, UK); and Ephemera (Ulises Carrión, Aart von Barneveld and Salvador Flores, Amsterdam, the Netherlands), to be discussed here, are key examples of the ways experimental art and ideas were exchanged within artist networks, while they also mark the increasing internationalism stimulated by the boom in printing and communications technologies. The circulation of such magazines bolstered the expansion of the mail art network, while also blurring the boundaries between the private and public aspects of mail communication. Considered together, the magazines also reflect the shared ethos of artists participating in the mail art network, and how they attempted to reinforce the principles of openness and egalitarianism upon which the movement was founded. The magazines were key in constructing the translocal communication engendered by the network, which in turn enabled the formation of a conceptual community between scattered groups of people who coalesced around common aesthetic interests and worldviews.

A precursor to mail art, Fluxus as a movement embraced publishing as an inherent challenge to the unique object and market-driven artistic production, bringing in multiply-produced "Fluxkits" such as George Brecht's *Water Yam* – which contained a multitude of cards with typeset event scores – that were conceived to be participatory and accessible.[3] Nam June Paik summed up the revolutionary nature of the development when he exclaimed, "Marx: Seize the production-medium. Fluxus: Seize the distribution-medium!"[4] An "assembling" is a publication for which artists send the entire run of their contribution to a coordinator, who is then responsible for its distribution.[5] Often, assemblings were unbound, constituting a collection of objects to be handled rather than mere pages to be turned. Assemblings were a drastic reconfiguration of the relationship between editors, contributors, and authors, in which artists took advantage of the availability of new printing methods in order to produce their own uncensored information. Importantly, the arrangement relieves the coordinators of the costs of printing, which are distributed amongst the participants. Assemblings became a key factor in constituting a networked community, exemplifying the shift from the periodical as a site for the *reproduction* of texts to one in which artist and reader collaborated in the *production* of new texts and experiences. As the addresses of the contributors were usually published as a routine element of the assembling, they became vehicles for consolidating and extending the network. Publications of this sort were often the instigator of a more direct, one-to-one contact since it was the primary way in which these new contacts could be found.

The collaborative nature of assemblings recalls the statement found on the cover of *The Blind Man*, published by Marcel Duchamp on April 10, 1917: "The second number of The Blind Man will appear as soon as YOU have sent sufficient material for it."[6] Many mail artists appropriated Dada strategies in their self-publishing, using bold design and copying techniques; asymmetrical print; random orientations; uneven letter sizes; satirical cartoons; adventurous typography; and cacophonous printed-over images or text. However, while the Dada artists did not conceive of their publications as artworks, mail artists produced "original-multiples" in which the artist had much more control over the outcome of the printing and dimensions than if the work was reproduced in a magazine.

Assemblings were inspired by, yet distinct from, Fluxus boxes and Fluxkits, which similarly could contain disparate works by a number of different artists. The initiator of Fluxus, George Maciunas, conceived of these Fluxus multiples as a challenge to the unique art object because they were made up of easily reproducible editions. However, in mail art the one-off work crept back into some assemblings circulated in the mail art network. If they so desired, an artist could feasibly produce the numbers required by hand. This created a tension between manual intimacy and the mass reproducibility that was enabled by the late twentieth-century boom in reproduction techniques. At one end of the spectrum, the artist could cheaply, quickly, and easily produce photocopies for maximum information distribution; at the other, they might create a handcrafted piece that aimed to connect with the recipient through its haptic qualities. Assemblings are therefore difficult to define, lingering somewhere between artists' books, artists' periodicals, and artworks.

This chapter discusses the four publications mentioned above that circulated through the international mail art network: Hexágono '71, Schmuck, Commonpress, and Ephemera. While not strictly speaking "assemblings," since most contributions were transformed by being printed by the editor, they are conceptually affiliated with the assembling, as we shall see. All conceived from unique points in the network, they are expressions of their editors' motivations and understanding of what a networked publication could achieve, and a desire to share points of view, correspondences, and collaborations with other artists. These publications are effectively a snapshot of the strategies and concerns of the artists who were drawn to collect, collate, and publish their correspondence. Their efforts allowed for the consolidation of ideas, reflecting the network back to itself while expanding links and connections between artists. The distinct vantage points of these magazines reveal the mobility of reception and dissemination enabled by publications produced through networking.

Hexágono '71

The Argentine artist Edgardo Antonio Vigo (1928–1997) approached his vanguard experiments through a synthesis of literary and mass culture, attempting a politicized practice that nevertheless retained connections with the artistic and literary avant-garde. Resident in the city of La Plata, at a 60-kilometer remove from the country's main artistic center, Buenos Aires, Vigo operated on the margins of contemporary art, in multiple senses. His oeuvre consists of conceptually oriented mail art, visual poetry, comic strips, performative actions, and expanded publications. From his quiet hometown, Vigo developed an extensive network of contacts, making the city a hub of the international experimental poetry and mail art network. While he explored ideas about mass media and alternative channels of communication in his work, Vigo nevertheless maintained an intimate and artisanal touch in his hand-printed works, which he described as cosas ("things"). Always concerned with the question of art's relationship with politics, he was strongly affected by political turmoil in Argentina, particularly after his young son was disappeared by the military junta in 1976. Vigo's mailed works were multiple but manifestly material, artisanal, and intimate "objects" or "things." Among other experiments with magazines, the artist spearheaded two substantial publications, *Diagonal Cero* and *Hexágono '71*. *Diagonal Cero* was an innovative, unbound magazine with object inserts developed within the international networks of experimental poetry. However, it was *Hexágono '71* that dates to Vigo's first explicit engagement with mail art.

Hexágono '71 (1971–1975) contains illustrations, visual poems, and works "to be realized": essays, drawings, stories, telegrams, and calls for submissions. In line with its engagement with experimental poetry and participation, *Hexágono '71* was composed of loose pages, which could be manipulated by the reader or "participant" and was distributed in a square envelope, which also acted as the publication's cover. *Hexágono '71* disposed with the conventions of publishing altogether, leaving out an editorial commentary, page numbers, editorial credits, location of the publisher, and the magazine's legal status, all of which were included in Vigo's previous magazine projects. Furthermore, *Hexágono*'s issues were ordered by a lettering system (rather than a numerical one): a, ab*, ac, b*c, b*d, b*e, cd, ce, cf, de, df, dg, and e.[7] The contributions were not stylistically associated in any way, and Vigo referred to himself as the "editor in-responsible" (a play between "nonresponsible editor" and "irresponsible editor"), adding, "There are no fixed collaborators, the issue takes shape as the works come in."[8] Indeed, the magazine did not operate on a fixed schedule; however, three issues per year were produced, with the exception of the final year, 1975, with only one.

The envelopes that stood in for covers were carefully made from heavy card (at the beginning, cream colored, and for the later issues, green) and were often perforated with holes, seemingly at random. These holes may indicate the instability in the concept of exterior/interior in a mail publication. However, the tension between public and private referred to by David Mayor here also relates to the double-edged issue of seeking out publics and the risk of drawing the attention of censors. Indeed, *Hexágono '71* began during the dictatorship of Juan Carlos Onganía and continued through the short-lived return of Juan Perón in 1973, followed by the disastrous term of Isabelita Perón, which led to the military coup of 1976.[9] This was a period of strong political censorship during which the tactics of the "dirty war" of 1976–1983 – summary executions, torture, and disappearances – were already becoming common.

Discourses around the international and national, regional, and local, arise in *Hexágono '71*: the magazine became increasingly international from 1971 to 1975, uniting artists from Argentina, Chile, Uruguay, Brazil, the U.S., France, Italy, and the UK.[10] Vigo's very specific articulation of the local was already present in his earlier magazine *Diagonal Cero*. In 1963, Vigo had forcefully stated *Diagonal Cero*'s local perspective: "being called localists does not bother us: WE ARE . . . without erecting barriers around national or foreign testimonies which are of vital importance for communicative progression."[11] *Diagonal Cero*'s valorization of locality was probably aimed just as much at Buenos Aires as at the international scene. But it is interesting that even in 1963, La Plata was conceived as a center from which "national or foreign testimonies" could flow unimpeded. Over ten years later, Vigo's idea was to "share the necessity of breaking the dangerous suffocation that hovers over the *universal* creative-investigator's free expression."[12] Indeed, issue ac of *Hexágono '71* published an essay by Fluxus affiliate Dick Higgins, and as such, was one of the first outlets for Fluxus ideas in Latin America.[13] This shift from the defiantly local to the ambitiously universal fluctuates throughout issues of *Hexágono '71*, which deal with concerns ranging from military dictatorship to media systems, and from artistic freedom to U.S. imperialism.

The covers of the first issues of *Hexágono '71* in 1971 and 1972 were inscribed with the phrase "UNO MÁS QUE USA" ["One more than the USA" or "One more who uses"] followed by the word "Punto" ["full-stop"], as if this statement were final. The strong anti-U.S. sentiment permeates *Hexágono '71*. Vigo's "Souvenir de Vietnam," a set of instructions in the first issue states, "Travel. Search for a trench (belonging to either side) gather a life and keep it in the transparent envelope?"; the envelope itself is provided. Another work, entitled "USA versus Latin American" appears in issue bc (Figures 4.1 and 4.2).

Figure 4.1 Edgardo Antonio Vigo, "USA versus Latin American," published in *Hexágono '71*, issue bc (La Plata, 1972).

Source: Image courtesy of Centro de Arte Experimental Vigo.

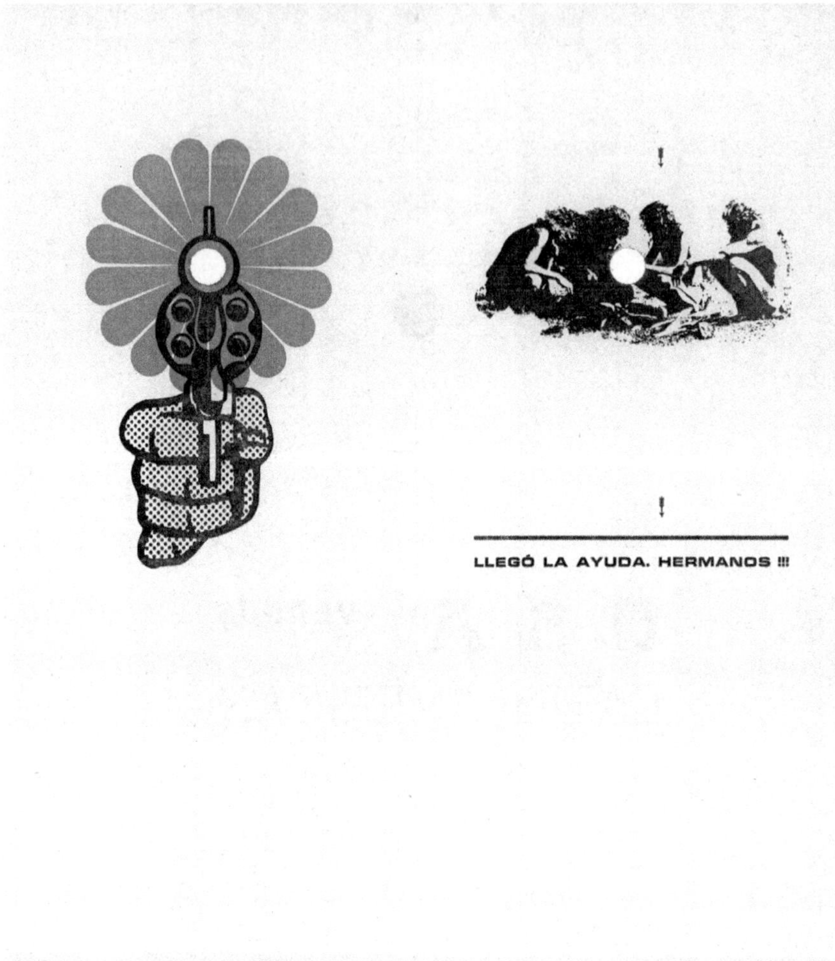

Figure 4.2 Edgardo Antonio Vigo, "USA versus Latin American," published in *Hexágono '71*, issue bc (La Plata, 1972).

Source: Image courtesy of Centro de Arte Experimental Vigo.

This time the hole suggests a bullet; when the page is turned, the reader is faced with two opposing images: the first is a gun, the hole forming a barrel pointing towards the viewer from which a red pattern emerges, which resembles a flower or an explosion rendered in the style of Roy Lichenstein; the second image is a group of Neanderthals gathered around a fire, under which is placed the caption "Llegó la ayuda hermanos!!!" ["The help has arrived, brothers!!!"]. This time the circular hole pierces the space in the center of the circle formed by the figures. The tone of Vigo's anti-US sentiment suggests a concern that transgresses merely contextual readings: On the one hand, it is a critique of US intervention in Argentina and

Latin America; on the other, his anti-Vietnam stance links him with the worldwide social revolution that was 1968.

Another negotiation and interchange of ideas was developed around the idea of Arte Povera, a movement in which the embrace of everyday "poor" materials definitively challenged the aestheticized object produced for the art market. The first issue of *Hexágono '71* contained a translation of a 1968 text written by Italian critic and curator Germano Celant. In it Celant argues for an object-based approach with an ethical commitment to the political processes of the day. By the final issue of *Hexágono* (e), in 1975, Vigo had formulated his own response to Celant's Arte Povera. Using the term "marginal communication," coined by French sociologist Hervé Fischer, he describes the specific marginality of the "creative Latin American research," against the "logical-traditional system." He also refers to the adverse economic situation in Argentina, which by 1975 was spiraling downhill:

> In Latin America, the REAL ARTE POVERA completely lacks state or private support, and when a representative confers it, it is always conditioned on a degree of absolute dependence. It is a kind of service, of which, we believe, is not avoided in Europe either or in any sphere of the researcher-creator, although, with different matrixes and percentages (their national panoramas).[14]

Here, Vigo acknowledges the artistic struggle for freedom of expression as a universal one, albeit inflected by varying circumstances.

By 1973, a shift in focus towards the "national reality" can be discerned. For issues cd, ce, and cf, the title was embellished with an image of Adam and Eve being driven out of Eden by a sword-wielding angel (Figure 4.3). Alongside the scene a statement reads, "eso sí, la más peligrosa" ["this yes, the most dangerous"], indicating that potentially volatile knowledge may be contained within. The stamps on the covers of issues cd and cf, "Argentine Vanguard Art" and "Investigation of National Reality," indicate a shift to a focus on national events; this is continued in 1974 with the cover of issue de stamped "Trelew," a reference to the 1972 massacre of sixteen political prisoners who were summarily executed after a failed mass escape attempt in Trelew, Patagonia. Issue df bears the stamp: "Libres o Muertos Jamás Esclavos" ["Free or Dead, Never Slaves"]. Issue dg is bound with a wide band of paper, across which is printed "Autocensurado" ["Self-censored"] (Figures 4.4 and 4.5). Ultimately, *Hexágono '71* was drawn to a close before the coup that saw the military return to power in 1976. It represents a snapshot of different approaches to experimental art, but also provides an account of Vigo's economic woes, anti-American sentiment and the political events occurring in Argentina at the time.

Like many in the mail art network, Vigo aimed to counter the disembodied bureaucracies of seemingly infinite reproducibility and the postal system by imbuing their works with a "live" aesthetic, one that presented the mailed work as an object of embodied materiality, and in which the paper artifact, archival document,

Figure 4.3 Edgardo Antonio Vigo, *Hexágono '71*, issue cf (La Plata, 1973).
Source: Image courtesy of Centro de Arte Experimental Vigo.

or ephemeral trace has the capacity to metaphorically "touch" a reader or recipient of a mailed work. Long-distance communication posed an intriguing problem for many mail artists; that is, while participation in the network usually signified a wholehearted adherence to a belief in the democratic potential of reproduction and a concomitant rejection of the "aura" of the work of art, artists also retained a need for the work's "magical" quality, since through this immediacy they were able to figuratively "touch" the recipient.[15] These references to corporeality are evident throughout many of the printed works that circulated through the mail system, achieved by means of overprinting, ink stains, and smudges.

While other mail artists (such as those I will go on to discuss in this chapter), often embraced a messy aesthetic full of traces of the human body that created the work, through various visible "errors," Vigo's work was, conversely, neat and pristine. However, there are at least two elements of it that require further examination in respect to their implication of the body. The first is Vigo's inclusion of irregularly shaped pieces of card in his mailed works or publications, which subtly

Figure 4.4 Edgardo Antonio Vigo, *Hexágono '71*, issue dg (La Plata, 1974).
Source: Image courtesy of Centro de Arte Experimental Vigo.

disrupt the clean lines and standardized formats of conventional publishing. The second is his life-long pursuit of a participatory aesthetic – Vigo's mailed works engaged the body by physically inviting participation. Indeed, the work was incomplete without the somatic-haptic collaboration of the recipient. Explicitly, in *Señalamiento VII de tu mano* ["Signaling VII: Of your hand"] included in the first issue of *Hexágono '71* (1971), for example, the page becomes animated by the instruction to insert one's hand into the five holes puncturing the square sheet of card. Indeed, Vigo's printed "things," as he called them, were non-normative

Figure 4.5 Edgardo Antonio Vigo, *Hexágono '71*, issue dg (La Plata, 1974).
Source: Image courtesy of Centro de Arte Experimental Vigo.

printed objects that invoked the object's phenomenological capacity. Thus the materiality of *Hexágono '71* acted as a conduit to a somatic, participatory encounter. While broad circulation was for him certainly an attractive prospect, as evidenced through the magazine projects Vigo spearheaded throughout his artistic career, it was also important to ensure the ethical consistency of the artwork that circulated. The non-normative objects and magazines that he circulated aimed to provoke a response far beyond mere participation, and instead provoke an "activation" in the viewer.

The Beau Geste Press: Schmuck[16]

One of the most active centers in the artists' book and mail art network was the Beau Geste Press, founded in 1972 in Collumpton, Devon, in the southwest of England. The Press's magazine *Schmuck* was published between 1972 and 1976 and was coordinated by English art historian David Mayor and Mexican artist Felipe Ehrenberg. Following the editorial principle – "the magazine makes itself, not we the magazine," corresponding to Vigo's policy of the "editor in-responsible" – Ehrenberg and Mayor allowed their Fluxus-influenced intermedial assemblage journal to be shaped by its contributors. Thus, each edition had a distinct flavor, and also (with the exception of *General Schmuck*) represented a snapshot of the alternative arts scene in each country represented (full issues were dedicated to Japan, Iceland, Hungary, Czechoslovakia, Poland, France, see Figures 4.6 and 4.7).[17] This approach also effected the timing of the release of issues, with long delays and some of the issues being released almost simultaneously. The editors started multiple conversations with their collaborators in each country and the magazines were produced according to all the participants' schedules, often with many delays.

As an alternative to one-way media monopolies, *Schmuck* enabled artists to disseminate information and engage in dialogue outside their own countries. This emphasis on developing community and democratic working practices outside of the dominant publishing institutions of the time reflects the aims of the wider mail art network. The title word, "Schmuck," a common insult meaning "penis" in Yiddish, continued the Press's taste for irreverent names, willfully contradicting the "beauty" of the "beau geste." *Schmuck* was not a mail art publication as such – indeed, David Mayor wrote to Taii Ashizawa and Ikuo Shukuzawa in 1973 asking for more material for the *Japanese Schmuck*, commenting: "I would really like to have some more 'solid' things from you than just mailed stuff, because I feel that it is difficult to print just-mail things in a magazine. . . ."[18] However, the magazine and its network did owe much to correspondence art and its networks, and this represents a substantial crossover among the participating artists. Many of the themes addressed by the artists, as well as the media in which they were expressed within the pages of *Schmuck*, were also central to mail art: bureaucracy, systems, language, concrete poetry, conceptual art, censorship, and politics. Moreover, Ken Friedman's correspondence networks had been crucial to the contributions to *Fluxshoe*, meaning that *Schmuck* was inextricably connected to the ideas and philosophy of the movement.[19]

Although based in the UK, the Beau Geste Press became a center for Latin American artists such as Ulises Carrión and Cecilia Vicuña, who either sent their work there or visited to print books themselves. The Press's publishing ventures deepened the casual structure of the assembling magazine by collaborating directly in the printing; because they functioned as an artists' book press, they were equipped to deal with the complexities of printing a magazine. In contrast to Vigo's magazines, the Beau Geste Press's communications, whether in the form of letters or books, enabled the development of affective relationships through an insistence on tactility and a certain aesthetic of cacophony – a "liveness"

Figure 4.6 Cover of *Schmuck* (1974).

Source: Cullompton, Devon: Beau Geste Press, 1974. Image courtesy of David Mayor.

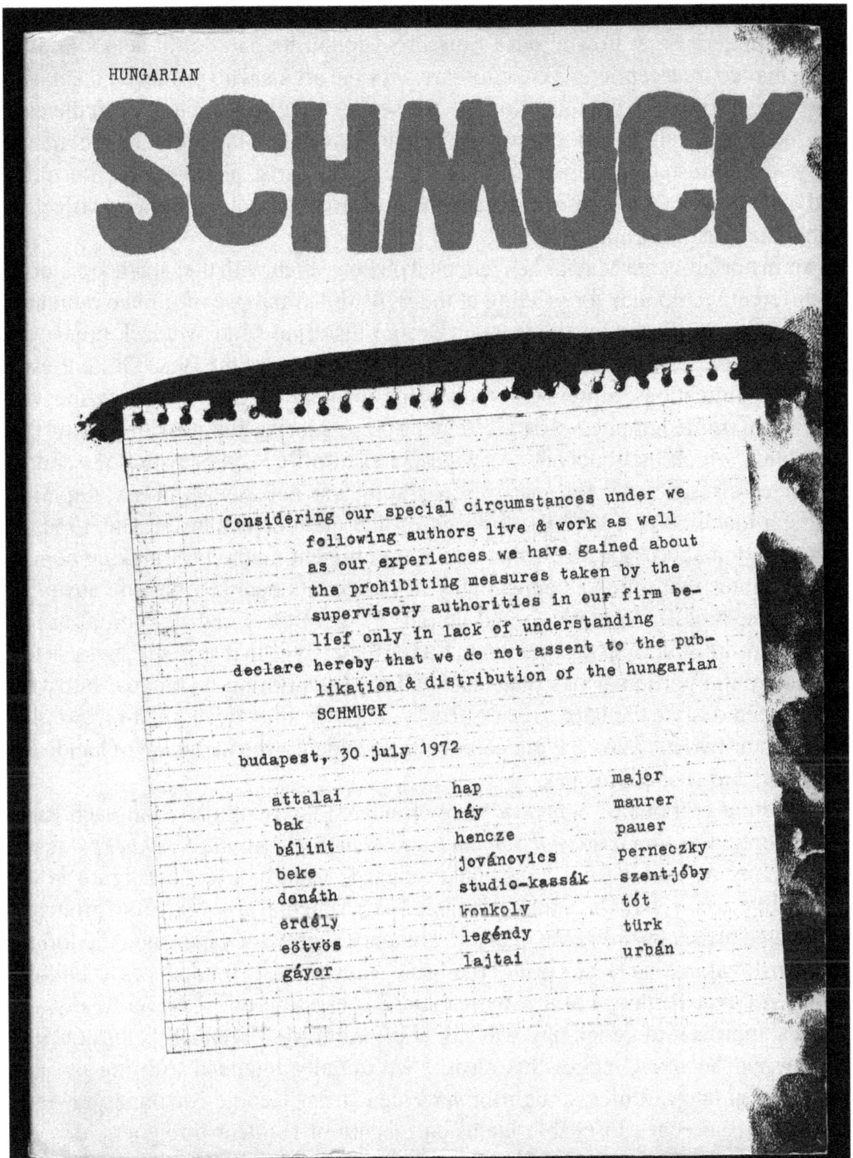

Figure 4.7 Cover of *Hungarian Schmuck* (1972).

Source: Cullompton, Devon: Beau Geste Press, 1972. Image courtesy of David Mayor.

that aimed to transcend the problem of spatial remove and effect the recipient or reader through the material qualities of the printed page. In some cases, the printed material that flowed out of the Press found its parallel in actual bodies; these mailed interventions were emissaries for the artists who produced them, and were transformed by the touch of both the sender and the recipient, or in the case of a book, the author and a newly participating reader. This index of the artist's body – manifested in fingerprints, outlined body parts, photographic portraits, and other bodily traces – is what made the mail art and book art network affective rather than just informational.

Art historian Vania Macias has reflected on how, even with the later acquisitions of different techniques for printing at the BGP, the initial use of mimeo remained an orienting aesthetic for the Press.[20] Design historian Mila Waldeck points out that in the magazine *Documento Trimestral* (1972), one of the Beau Geste Press's earliest publications, Ehrenberg took pains to mention that the magazine was "carefully mimeographed on a Gestetner 300, exploring the possibilities of this medium," which had become "so widespread in offices and revolutions and so disdained elsewhere."[21] Because mimeography was notoriously messy, the mimeograph machine was broadly replaced by the photocopier in the late 1960s, a technology that provided a cleaner and more efficient method for making copies. Macias notes that mimeo's appeal may have been its connotations of "austerity, resistance, protest, irreverence and utopia."[22] She argues that this initial use of mimeo in the first *Schmuck* generated an artistic style that initially "was determined by the particularities that belonged to that printing technique, but were maintained despite the later use of offset" – methods and effects such as "printing with one or two inks, overlaying colours, the accidents, the inclusion of handwritten texts, and stained finishes."[23]

The eight editions of *Schmuck* were circulated internationally but each (with the exception of two, *General Schmuck* and *General Teutonic Schmuck*) was put together by an editor from a particular country, usually a leading figure in the alternative arts scene. The editions included experimental artworks from groups of associated artists from France, Iceland, Hungary, Germany, Japan, and the former Czechoslovakia. The Beau Geste Press was "viewed by its founders as a 'link-up' between Great Britain, Latin America and Eastern Europe."[24] Nevertheless, the Press's approach to geography was not at all arbitrary. Ehrenberg's introduction to *General Schmuck* makes this clear: "We initially intended to bring out one issue per country, a nice straightforward idea (from George Maciunas), woolly enough at the edges to enable inclusion of current resident foreigners, or non-resident nationals, in my particular case, plurinationalism."[25] These comments reveal flexibility in relation to geography that undermines fixed ideas of place and nation. *Schmuck* was a publication that practiced internationalism through specific articulations of the local.

Schmuck was at least partly informed and inspired by *El Corno Emplumado/The Plumed Horn*, an independent publication founded by Margaret Randall, Sergio Mondragón, and Harvey Wolin in Mexico City. Ehrenberg contributed to several issues, which also marked his first experience of working in a collective and independent publishing venture. Bilingual, experimental, and collaborative in nature, *El Corno Emplumado/The Plumed Horn* published Beat poets as well as key figures in poetry from Latin America, such as the Brazilian concrete poet Haroldo de Campos.[26] Ehrenberg described it as "a highly politicized and far-reaching magazine which gathered the most prominent and meaningful poets, writers and artists in Anglo and Latin America."[27] However, by 1969, Randall and Mondragón were forced by the Mexican government to close down the publication. The collective experience, and disproportionate political reaction to it, demonstrated the rich artistic potential of publishing with other artists and poets, while simultaneously making it clear that limiting the circulation and freedom of information was crucial to the government's attempts to maintain the status quo. Having control of a means of producing and disseminating information thus appeared to be a highly political act.[28] Recalling it as a direct response to the repressive atmosphere, the artist and publisher Marta Hellion remembers that the magazine was "the first opportunity to reflect and compile works, ideas and political situations in a magazine: for the first time we became aware of Latin America."[29]

Central to the success of the Beau Geste Press was interaction and participation with "as many possible sources" of "small groups of creators" that have "something unnameable in common."[30] The magazine *Schmuck* is exemplary in this respect. One of the key aspects of Beau Geste Press production is that through *Schmuck* the Press established contacts with groups of artists, often operating collectively, in order to foster and consolidate a parallel system that could function outside of the "monopolic control of culture."[31] *Schmuck* provides a record of the connections, collaborations, and convictions of the Press and reflects the strategic and pragmatic changes made during that period. It was an assembling magazine composed of printed multiples and object inserts by the contributing authors and artists, and it brought together a variety of artists from a range of geographical spheres.[32] A "network of encounters and exchanges,"[33] its circulation helped disseminate new ideas about art making. Both the art presented and the essays that appeared within its pages aimed to produce and spread alternative ideas and knowledge. Independent presses hoped to demonstrate, in Felipe Ehrenberg's words, "how easy and viable it is to ignore publishers and producer-galleries."[34] The publication was intimately connected with the rejection of traditional art spaces and a nonhierarchical approach to the production and consumption of art, surveying the "then diffuse terrain of nonconformist art," and explicitly devoting itself to distancing itself from the "speculative glitter" of the mainstream art world.[35]

Ephemera

In November 1977, Ulises Carrión, Aart von Barneveld, and Salvador Flores put out the first issue of *Ephemera* magazine, "a monthly journal of mail and ephemeral art," as it was stated on the front cover of the slim publication. The magazine was a compendium of works received through the mail, which were subsequently pasted up, and reproduced in a simple broadsheet format. The quick turn-around of the magazine is indicative of the sheer amount of material being received by the artists on a monthly basis, using a quick method of paste-up and print, likely with a mimeo press. The front and back cover were adorned with the slogans of participants in the mail art network: "ART SPOKEN HERE" (Judith Hoffberg); "mail art it's fine art" (a.ferro et al.); "do as you like R.F" (Robert Filliou); "EACH DAY TO COME FOR ONE YEAR MORE – ROY GRAYSON, LONDON." Mail stickers, rubber stamps and photocopied mailings were scattered across the page, all neatly annotated with Carrión's handwriting detailing artist's name, title, date, and the work's place of origin, in a distinctly Dada-inflected approach to typography. The pages are mostly black and white, but are scattered with color stamps and mail stickers, and Carrión's distinctive cursive gives the issue a personal touch. Prominent on the front cover, a photograph of a jar full of fingernail clippings by an artist identified only as "Alessandro" is dated 1971–1976, indicating the jar contained half a decade's worth of nail clippings, presumably the artists own (Figure 4.8).[36]

Among many other reproduced items from Carrión's mailbox is *Postman's Choice*, a postcard created by Ben Vautier in 1965. The postcard has an address section on both sides, enabling the sender to fill in two addresses: the postman is to choose which to send it to. In *Ephemera*, it is shown in its sent form, with "Ulises & Art" being the eventual recipient, and "Beste Helen" losing out. Carrión annotates the reproduced piece: "We (Ulises & Art) got it," as if they had won a prize. The following pages continue in the same vein with contributions from artists as far flung as Dick Higgins, Richard Prince, Gábor Tóth, and Paulo Bruscky. The editors succinctly cite other mail art publications that have inspired their own:

> We like Diego Barbosa's "Arte Correo."
> We like Raúl Marroquin's "Fandango."
> We like Anna Banana's "Vile."
> We like Klaus Groh's "Info."

All of the above cited publications provided news and digest-style assembling compilations to the mail art network, with differing editorial perspectives. The deliberate choice to name them in *Ephemera* further indicates the ways in which mail art networks were propagated – not only through a postal exchange and the printing of the resulting artifacts, but also by directly naming other publications of influence, and ostensibly encouraging readers to take note of the work of the artists' peers.

Later issues of *Ephemera* continue in the same vein. Each of the magazines incorporated additional elements or distinct formats that elevated the publication from a mere compilation of reproduced sheets, and thus indicated it was carefully handmade. *Ephemera 6* for example, was adorned with brown paper receipts sent by Robin Crozier. This issue was headlined by Gábor Tóth's "sorry no art today." *Ephemera 5* was notable in that, in order to be opened, it had to be cut open along a dotted line, an action proposed by the inclusion of a scissor icon. This format mischievously blurs the line between innovative envelope, censored information, and playful participation. On the front page of the same issue, the exclamation "Padín and Caraballo are in Jail!" was also incorporated into the editor's design, and the statement is annotated, in Carrión's familiar handwriting, with the remark,

Figure 4.8 Cover of *Ephemera* 1 (Amsterdam, November 1977).

Source: Image courtesy of Estate of Ulises Carrión.

"Ask for information. Transmit it," thus bringing forth concerns about freedom of information, as well as an active, participatory engagement with media and information distribution (Figure 4.9). The final two issues were dedicated to specific countries: the "Hungary Special," issue 11, is almost a poster, in that each work is subordinated to an overall design schema and "Ephemera Brazil," issue 12, consists of mailed pieces from people associated with the experimental poetry movement *Poema/Processo* (Poetry/Process) that aimed to bring poetic discourse into the performative and participatory realm. Works from Latin America and Eastern Europe were hardly underrepresented in the rest of *Ephemera*, as librarian and researcher Sarah Hamerman has noted. Nevertheless, she points out that

> on the one hand, the Hungary and Brazil issues serve to showcase the aesthetic specificity and disciplinary origins of mail art practices within these regions. On the other, by showcasing mail art practices in local contexts outside western democracies, *Ephemera* deepened artistic exchange across geographic and ideological borders, making a small but significant political intervention (Figure 4.10).[37]

Guy Schraenen has noted that *Ephemera* reveals "an immense diversity in the aesthetics, conception, and geographical origins of the works," with each participant "forged into a collective work."[38] The mail art network itself was a multiauthored conceptual project made of hundreds of thousands of works. Cutting, pasting, and collaging these disparate works together formulates the aesthetic of the network, of multivocal, interdialogic, and process-based character, that likewise gestures to its participation in the larger network of mail art publications, such as Beau Geste's *Schmuck*. Indeed, around the time of the publication of *Ephemera* magazine, Carrión wrote, "From Bookworks to Mailworks" (1979), in which he explored the question of authorship in mail art: "In a project containing 150 pieces, am I to be considered the author of only that one showing my signature? Am I innocent of the other 149? All 150 pieces should rather be considered as 'one' element in a complex artwork."[39] For Carrión, the multiauthored work indicates a shift from "artwork" to "cultural strategy," which also encompassed his daily activity reproducing and distributing mail art. He observed that

> Mail Art shifts the focus from what is traditionally called 'art' to the wider concept of 'culture.' And this shift is what makes mail art truly contemporary. In opposition to 'personal worlds,' mail art emphasizes cultural strategies. This radical shift gives birth to quite a number of theoretical and practical questions, the most evident of them being, where does the border lie between artist's work and the actual organization and distribution of the work?[40]

The way in which the magazine *Ephemera* is collated from individual works circulating in the mail art network is an analogy for the network itself. The magazine becomes, through collaborative collage, a multiauthored work that is pasted together to form something new.

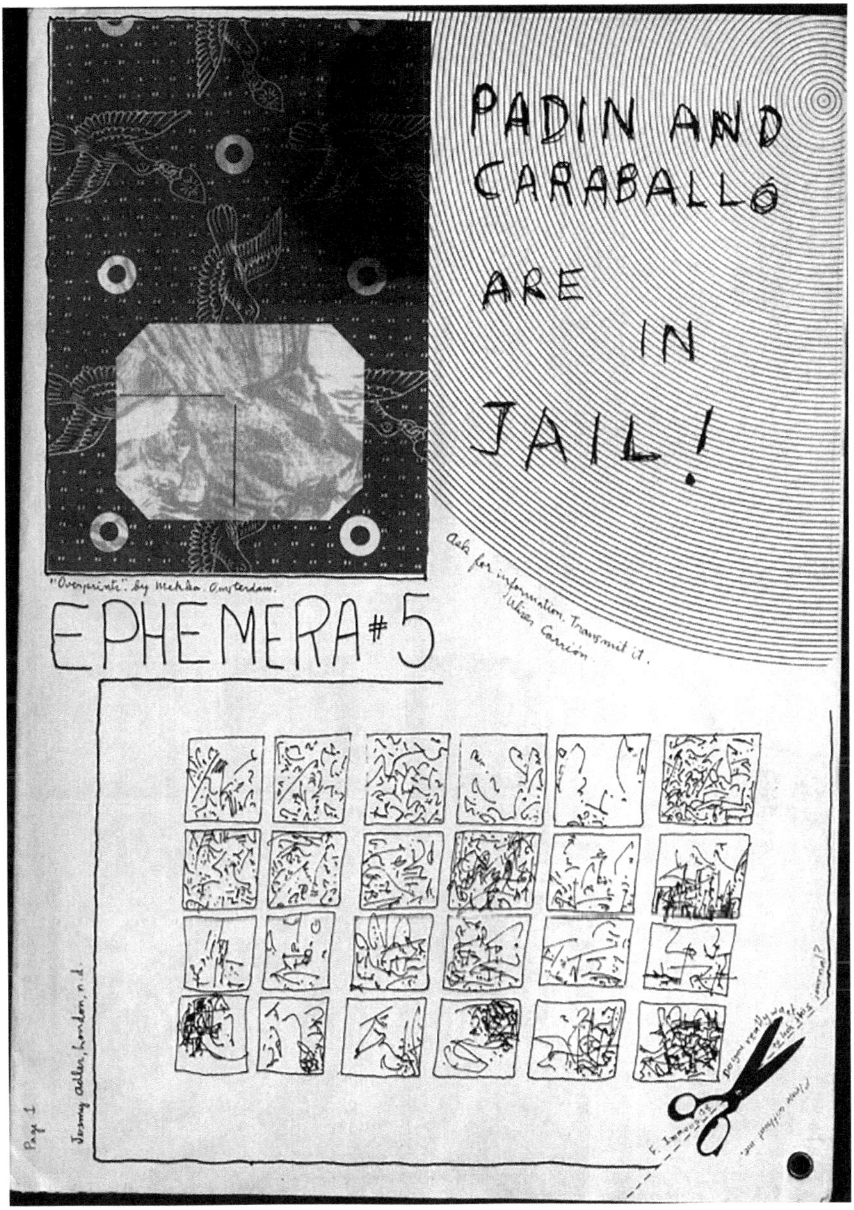

Figure 4.9 Cover of *Ephemera* 6 (Amsterdam, April 1978).

Source: Image courtesy of Estate of Ulises Carrión.

Figure 4.10 Cover of *Ephemera* 12: *Brazil* (Amsterdam, October 1978).
Source: Image courtesy of Estate of Ulises Carrión.

Paulo Bruscky & Leonhard Frank Duch's Commonpress*: "Post Office"*

While Edgardo Antonio Vigo and the Beau Geste Press both proclaimed their editing to be extremely hands off, and *Ephemera* had more of a personal touch, ultimately the apex of the collapse of traditional roles in publishing is marked by Pawel Petasz's project *Commonpress*, which relied entirely on the mail art network to produce rotating editors. *Commonpress* (1977–1990) aimed to be exactly what its title proclaimed: a collectively undertaken publishing venture supported by an international network of collaborators. It was established by Petasz, an artist from Poland, with editorial responsibility being passed on like "a relay baton."[41] Géza Perneckzcky explains how the system functioned:

> The paper had a central coordinator, who assigned each issue to an editor. The editors chose a subject, and announced the proposed format and editorial deadline. Since *Commonpress* was agreed to be issued in 200 copies only, the contributing artists were advised to submit their works in the same number of copies. Each was obliged to announce the addresses of the designated editors of the next issue.[42]

Perkins points out that Petasz's "innovation was to use the international network as surrogate editor and compiler of this magazine."[43] The rotating editorial responsibilities acted as a kind of chain reaction and Petasz eventually ceded all control due to the difficulties of the political situation in communist Poland, which limited the possibilities for independent publishing.[44] This "perpetual periodical machine" produced sixty-one issues in the thirteen years that it was running, a period that long outstrips other mail art publications and that vindicates its collaborative method.[45] Sometimes a batch of *Commonpress* issues happened to be distributed at the same time, the excess of information functioning like the result of a glitch in an overloaded computer system which spews out random nonsensical code. However, within this glut of information exists the chaos that can be taken as a source of possibility in an open and heterogeneous network. When the Canadian artist Jupitter-Larsen took over *Commonpress* from Petasz, he defined the project "not as an alternative magazine but an expanding international performance."[46] He referred to a plethora of perspectives and a process-based, performative approach that is reflected in the *Commonpress* logo – a set of theatrical curtains with the periodical's title inscribed across them.

If the assembling model made it possible for a coordinator to produce a publication at low cost but demanded a great deal of energy, Petasz's idea set the coordinator free from an ongoing burden. In a flyer soliciting works for the first issue of the magazine, Petasz states:

> *Commonpress* is a conception of the periodical edited by common effort. Possible realisation of this conception would be let to overcome such difficulties as print and distribution expenses, nothing to say about the dangers of commercialisation [*sic*].[47]

The question that arises, given the extreme freedom ceded to each editor, is why each collaborator did not simply undertake individual publishing projects under different names? The reason lies in the fact that *Commonpress* represented a ready-built network with an already proven formula, and an opportunity to be part of something greater than an individual project, to participate in a wider, previously established network. Perneckszky notes that, "after a while the regularly contributing artists became members of a kind of community, and in that spirit, they can cooperate almost flawlessly."[48] Perkins argues that *Commonpress* is a "meta-assembling" only possible because mail art became a self-sustaining network.[49] Indeed, one might consider the behavior associated with the publication to be more important than what was actually published, affirming the conceptual and anti-market principles of process over product: the network *is* the work.

If the network itself is what is being created and consolidated through the mail art assembling, this was undoubtedly reflected in the tenth issue of *Commonpress*, organized by Paulo Bruscky and Leonhard Frank Duch in 1978. The issue was unique in that the theme chosen by the editors was "Post Office." As a result, the issue is one of the most concentrated examples of mail art's self-referential character: the medium is quite literally the message. In "Post Office," the network itself and the artists' method of collaboration are explored from varied perspectives. The issue contains an extraordinary variety of meditations on the theme: mailed envelopes; address labels; official postal paraphernalia; parodic applications for employment with the postal service; images of postmen; makeshift imaginary post offices; sketches of postal workers in the sorting office looking at mail art (Petasz's "What Art? Mail Art?"); instructions to "hug a postman"; diagrams of postal bureaucracies; pornographic allusions to "slots"; instructions to gather artwork from postal clerks; self-made stamps; and Carrión's instructions for the use of the Erratic Mail Art International System (E.A.M.I.S), of which he was the Post Master. Carrión's proposal was an alternative postal system that delivered "messages in any format . . . by any way other than the official Post Office."[50]

The contributions range from the hilarious to the subtly beautiful or gravely touching. Italian contributor Augusto Concato sent in a contribution titled "Myself to Myself," a photograph of the artist packaged for mailing and ready to be sent to himself; the conceptual tautology providing an added laugh. An elegant drawing by G. E. Marx Vigo (the collective pseudonym of *Hexágono '71* editor Edgardo Antonio Vigo and his collaborator Graciela Gutiérrez Marx) of four hands connected by chaotic and loosely woven threads is accompanied by the stamp "Seen and heard" and, across a torn envelope, their fragmented address. The hands appear to be deformed and unable to grasp or control the network of threads that bind them. These threads perhaps allude to the censorship of the artists' mail – their "hands are tied" – but perhaps equally they point to a residual hope in the capacity of the network to support artists. The grace of this image is accompanied by a sense of fragility in terms of the connection to a network in a period of grave crisis in Argentina – the stamped statement alludes to the ever-present possibility of disappearance and the use of the network to affirm and record one's presence. This work was sent shortly after the disappearance of Vigo's young son, Abel

Palomo, in July 1977.[51] *Commonpress* attempts to transcend cartographic boundaries through its itinerant structure, coming close to the locally inflected universality that *Hexágono '71*, *Ephemera*, and *Schmuck* aimed for. Nevertheless, works like the one by G. E. Marx-Vigo provide glimpses of the local situations they at once transcended and reflected.

Curiously, however, it seems as though the network of *Commonpress* 10 was conjured but not fully unleashed as the magazine was apparently never published or distributed, according to Leonhard Frank Duch.[52] However, one version exists in the archive of Paulo Bruscky in Recife and Duch has an alternative version in his own archive in Berlin. *Commonpress* offered an alternative publishing model, but one without editorial oversight, resulting in glitches in the system such as unpublished issues, or in this case multiple versions of an issue.

The magazines *Hexagono '71*, *Schmuck*, *Ephemera*, and *Commonpress* were based on four distinct assembling formats that differed in how materials were gathered, printed, and distributed. Editorial control ranged from Vigo's tight oversight of his publication, to the collaborative efforts of the Beau Geste Press's distributed network of collaborators, the collectively controlled, self-sustaining *Commonpress*, and the lovingly annotated *Ephemera*. While *Hexágono '71* was carefully and professionally printed, the Beau Geste Press used their own well-equipped printing press studio, *Ephemera* was a more DIY job, and the works included in *Commonpress* were often photocopied. Each of the loosely defined "editors" act as collator and duplicator, providing the network feedback to itself. Writing about punk in East and West Germany, author Christian Schmidt reflects that shared values "must find material expression in shared practices. . . . Only through these practices can common attitudes become manifest, so those inside and outside of the scene can notice them and it can be understood a scene exists at all."[53] This function of community-creating can help to explain why so many assembling-style magazines proliferated through the network. Another shared characteristic is that of plurality; no one magazine could serve the needs of the network, since the network itself is characterized by its many distinct modes of "being together." Furthermore, all four magazines shared the common aims of distributing uncensored information across vast distances, as such they provide an early account of an incipient globalization of art networks and discussions of translation between locales, of the condition of being at once "here" and elsewhere. All of the magazines engage with political situations and events in ways ranging from the oblique to the direct – freedom of movement is a particularly apt subject matter for the artists who contributed to these magazines. These self-organized publications pushed publishing conventions to their limits, exacerbating the performative characteristics of networked art.

Notes

1 An earlier version of this essay was previously published in *A+WS: A propósito de nada / Apropos of nothing* (Mexico City: Aeromoto/Wendy's Subway, 2017), a collaborative publication produced by the two libraries Wendy's Subway (Brooklyn, New York) and Aeromoto (Mexico City, DF). This earlier version was also published in

Spanish on the Museo El Eco website as "Redes de Impresión: Publicaciones colaborativas y la red de Arte Correo," accessed April 30, 2018, http://eleco.unam.mx/eleco/redes-de-impresion-publicaciones-colaborativas-y-la-red-de-arte-correo/.

2 David Mayor, "Introduction," in *Schmuck* (Devon, South Cullompton: Beau Geste Press, 1972), 2.

3 See Jon Hendricks, ed., *Fluxus Codex* (New York: Harry N. Abrams, Inc., in association with The Gilbert and Lila Silverman Fluxus Collection, Detroit, MI, 1988); Hannah Higgins, *Fluxus Experience* (Berkeley, CA: University of California Press, 2002); Ken Friedman, ed., *The Fluxus Reader* (New York, NY: Academy Editions, 1998).

4 Nam June Paik cited by Craig J. Saper, "Intimate Bureaucracies and Infrastructuralism: A Networked Introduction to Assemblings," *Postmodern Culture* (May 1997): 6.

5 See Craig J. Saper's, *Networked Art* (Minneapolis: University of Minnesota Press, 2001); Stephen Perkin's, "Artists' Periodicals and Alternative Artists' Networks: 1963–1977" (PhD diss., University of Iowa, May 2003) for more information on the phenomena of the assembling magazine.

6 Marcel Duchamp, *The Blind Man* no. 1 (1917).

7 The letters correspond to years: a (1971), b (1972), c (1973), d (1974), e (1975). The letters are always followed by another letter that comes after that letter in the alphabet. The "b" is always starred, but the reason for this is unclear to me, nor have other scholars commented. Three issues were published per year, except in 1975, when only one issue was realized.

8 Author anon, "*Hexágono U.N.O más que U.S.A*," *El Día*, Sunday Supplement, La Plata, January 7th, 1973. ["Colaboradores fijos no hay: el número se va armando a medida que llegan los trabajos."] Cited by Vanessa Davidson, "Paulo Bruscky and Edgardo Antonio Vigo: Pioneers in Alternative Communication Networks, Conceptualism, and Performance (1960s–1980s)" (PhD diss., New York University, Institute of the Fine Arts, 2011), accessed July 27, 2012, http://gradworks.umi.com/3482872.pdf.

9 *Hexágono '71* ceased publication before the worst years of oppression in Argentina, and so it must be assumed that – whilst censorship was a real issue under Onganía – the production of magazine like *Hexágono '71* was not as risky as it would become after 1976.

10 These artists included Jochen Gerz, Carlos Ginzburg, and Klaus Groh, among many others.

11 Edgardo Antonio Vigo, ed., *Diagonal Cero*, vol. 7 (1963) ["No nos importa si nos llamen de localistas: LO SOMOS. [. . .] *DIAGONAL CERO*, sin poner barreras a los testimonios nacionales y extranjeros de vital importancia para el progreso comunicativo."]

12 Carlos Gradin, "Un foco en la selva de los medios. Revista *Hexágono* (1971–1975)," *revista laboratorio* no. 4 (2011): 4, accessed April 6, 2012, www.revistalaboratorio.cl/2011/06/un-foco-en-la-selva-de-los-medios-revista-hexagono-1971-1975/.

13 Ibid., 2. I consider mail art an unorthodox offspring of Fluxus, but much broader in scope, though tangentially related and following some similar principals. There were also several crossover members, such as Ken Friedman, mail artist and Director of Fluxus West, as appointed by Maciunas. For more details on the relationship between mail art, Fluxus, Conceptualism, and experimental poetry, see: Zanna Gilbert, "Genealogical Diversions: Experimental Poetry Networks, Mail Art and Conceptualisms," *Caiana* 4 (2014), http://caiana.caia.org.ar/template/caiana.php?pag=articles/article_2.php&obj=136&vol=4.

14 Edgardo Antonio Vigo, "Una forma real de arte pobre," *Hexágono '71* no. e (1975): unpaginated. ["En Latinoamérica el REAL ARTE POBRE carece de total apoyo estatal o privado, y cuando algún ente representative lo confiere, siempre está condicionado al grado de dependencia absoluta. Es un estado de servicio, del cual, pensamos, tampoco escapan en Europa o en cualquier lugar del orbe el investigador-creativo, aunque, con matices y porcentuales diferentes (sus panoramas nacionales). Quizás."]

15 Ian Walker, *City Gorged with Dreams: Surreal Documentary Photography in Interwar Paris* (Manchester: Manchester University Press, 2002), 12.
16 For an expanded version of this section, and a more detailed analysis of Schmuck, in particular the relationship of the Beau Geste Press with Eastern Europe and Latin America, see Zanna Gilbert, " 'Something Unnameable in Common': Translocal Collaboration at the Beau Geste Press," *ARTMargins* 1, no. 2–3 (2012): 45–72.
17 If it seems curious that there was no "Mexican" or "Latin American Schmuck," especially since Ehrenberg was from Mexico, there was a plan for an issue that focused on the continent, and indeed, submissions were solicited and sent for this issue. However, the project was never produced as an issue of *Schmuck* and instead took on another life once Ehrenberg returned to Mexico City. For a full account of the "missing Schmuck" see: Gilbert, " 'Something Unnameable in Common': Translocal Collaboration at the Beau Geste Press."
18 Letter from David Mayor to Taii Ashizawa and Ikuo Shukuzawa April 1973, David Mayor Collection, Tate Archive.
19 *Schmuck* was "based on an increasingly expanded network of mail interchanges by artists carried out in the tradition of La Monte Young's *An Anthology* (1961) and Maciunas' *Flux Yearboxes* (1962)." [Olivier Debroise and Cuahtémoc Medina, *La Era de la Discrepancia: arte y cultura visual en México 1968–1997/The Age of Discrepancies art and visual culture in Mexico* (New York and Mexico City: Turner and UNAM, 2007), 158].
20 Vania Macias Osorno, "Schmuck y la Beau Geste Press: Todos somos uno," *Reflexiones Marginales* 6, no. 41 (October–November 2017): 9, http://reflexionesmarginales.com/3.0/schmuck-y-la-beau-geste-press-todos-somos-uno/.
21 Ehrenberg, quoted in Mila Waldeck, "Perfect Unbinding: The Production and Circulation of Beau Geste Press Editions," *Journal of Artists Books* no. 41 (Spring 2017): 8.
22 Ibid.
23 Osorno, "Schmuck y la Beau Geste Press."
24 Ibid.
25 Felipe Ehrenberg, *General Schmuck* (South Cullompton, Devon: Beau Geste Press, 1975), 1.
26 Many of the Beat generation poets, including Margaret Randall, spent time in Mexico from the 1950s onwards, as did Latin American poets such as Ernesto Cardenal, Raquel Jodorowsky, and Roger Bartra. Robert Creely, Rothenberg, Nicanor Parra, William Carlos Williams, Cecilia Vicuña, and Philip Lamantia were some of those who contributed to *The Plumed Horn*. See Zanna Gilbert, "The Eclectic World of Felipe Ehrenberg," *ESTRO* no. 2 (2010): 48
27 Quoted in Gilbert, "Eclectic World," 48.
28 The history of the printing press in Mexico is intricately connected to the Mexican revolution and the following years of unrest as the new establishment attempted to consolidate power. Ehrenberg also spent time in a printing workshop run by Catalonian anarchists, who were exiled in Mexico during Franco's rule. [Author's conversation with the artist, December 2009.]
29 Martha Hellion, "Artists Books from Latin America," *Printed Matter: Critical Essays*, November 2006, www.printedmatter.org/researchroom/essays/latin.cfm.
30 Felipe Ehrenberg, writing under the pseudonym Kyosan Bajin in *Schmuck*, quotes George Brecht, who said in 1964, "Individuals with Something Unnameable in Common Have Simply Naturally Coalesced to Publish and Perform Their Work," in *Schmuck* (South Cullompton: Beau Geste Press, 1972), 1.
31 Ehrenberg quoted in Debroise and Medina, *La Era de la Discrepancia*, 158.
32 For more information on assembling magazines, see Craig Saper, "Intimate Bureaucracies and Infrastructuralism."
33 It was described as such by Debroise and Medina, *La Era de la discrepancia*, 157.
34 Ehrenberg, "Introduction," 2.

35 Debroise and Medina, *La Era de la Discrepancia*, 158.
36 This work is likely by Italian mail artist Alessandra Ceccotto.
37 Sarah E. Hamerman, "Beyond Bookworks: Ulises Carrión's Cultural Strategies" (MA Thesis, Pratt Institute, 2017): 39.
38 Guy Schranean, *Ulises Carrión, Dear Reader: Don't Read* (Madrid: Museo Nacional Centro de Arte Reina Sofía, 2015), 22.
39 Ulises Carrión, "Personal Worlds and Cultural Strategies," 1979, www.artpool.hu/MailArt/chrono/1979/Carrion.html.
40 Ibid.
41 Géza Perneczky, *The Magazine Network*, trans. Tibor Szendrei (Koln: Soft Geometry, 1991), 15.
42 Ibid.
43 Perkins, "Artists' Periodicals and Alternative Artists' Networks," 138.
44 Perneczky, *The Magazine Network*, 15.
45 Perkins, "Artists' Periodicals and Alternative Artists' Networks," 139.
46 Quoted by Perneczky, *The Magazine Network*, 15.
47 Quoted by Perkins, "Artists' Periodicals and Alternative Artists' Networks," 138.
48 Perneczky, *The Magazine Network*, 16.
49 Perkins, "Artists' Periodicals and Alternative Artists' Networks," 138.
50 Reproduced in Mauricio Marcin, *Arte Correo en un Libro* (Mexico City: Museo de la Ciudad, 2011), 246.
51 Vigo describes the event in a text found in his archive held at Centro Experimental Vigo, La Plata, Buenos Aires.
52 Conversation with the author. Berlin, May 19, 2018.
53 Christian Schmidt, "Meanings of Fanzines in the Beginning of Punk in the GDR & FRG," *Volume!* 5, no. 1 (2006), 47–72, 59.

Bibliography

Bruscky, Paulo. "Arte Correio e a Grande Rede: Hoje, a Arte é este Comunicado." (1976, reworked 1981), Canal Contemporâneo. www.canalcontemporaneo.art.br/blog/archives/004232.html.

Bruscky, Paulo. "Xerografía artística: arte sem original (da invenção da maquina ao processo Xero/gráfico)." In *Arte Novos Meios/Multimeios-Brasil, 70/80*, edited by Daisy Peccinni. São Paulo: FAAP, 1985.

Carrión, Ulises. "Personal Worlds and Cultural Strategies." 1979. www.artpool.hu/MailArt/chrono/1979/Carrion.html.

Cook, Geoffrey. "The Padín/Caraballo Project." In *Correspondence Art*, edited by Michael Crane, 369–373. San Francisco: Contemporary Art Press, 1984.

Cook, Geoffrey. *Umbrella* 2, no. 1 (1979): 15.

Crane, Michael, and Mary Stofflet, eds. *Correspondence Art: Source Book for the Network of International Postal Art Activity*. San Francisco: Contemporary Arts Press, 1984.

Davidson, Vanessa. "Mailings from the Margins: Paulo Bruscky, Leonhard Frank Duch and Edgardo Antonio Vigo's Mail Art Practice." https://sites.google.com/a/artproof.co.uk/art-proof/home/journal/issues/issue2/.

Davidson, Vanessa. "Paulo Bruscky and Edgardo Antonio Vigo: Pioneers in Alternative Communication Networks, Conceptualism, and Performance (1960s–1980s)." PhD diss., New York University, Institute of the Fine Arts, 2011, accessed July 27, 2012. http://gradworks.umi.com/3482872.pdf.

Davis, Fernando. "Prácticas 'revulsivas'. Edgardo Antonio Vigo en los márgenes del conceptualismo." In *Conceitualismos do sul/Conceptualismos del sur*, edited by Cristina Freire and Ana Longoni. São Paulo: Annablume, 2009.

Davis, Fernando. "Señalar y revulsionar. Edgardo Antonio Vigo en los márgenes de la poesia." In *V Jornadas sobre Arte y Arquitectura en Argentina*. Buenos Aires: UNLP, 2007.

Debroise, Olivier, and Cuahtémoc Medina. *La Era de la Discrepancia: arte y cultura visual en México 1968–1997/The Age of Discrepancies Art and Visual Culture in Mexico*. New York and Mexico City: Turner/UNAM, 2007.

Freire, Cristina, and Ana Longoni, eds. *Conceitualismos do Sul/Conceptualismos del Sur*. São Paulo: Annablume, 2009.

Friedman, Ken, ed. *The Fluxus Reader*. Chichester: Academy Editions, 1999.

Friedman, Ken in Ronnie Cohen, ed. "Mail Art History: The Fluxus Factor." *Franklin Furnace Flue* 4 (Winter 1984): 18–24.

Gilbert, Zanna. "Genealogical Diversions: Experimental Poetry Networks, Mail Art and Conceptualisms." *Caiana* 4 (2014). http://caiana.caia.org.ar/template/caiana.php?pag=articles/article_2.php&obj=136&vol=4.

Gilbert, Zanna. "'Something Unnameable in Common': Translocal Collaboration at the Beau Geste Press." *Art Margins* 1, no. 2–3. Cambridge, MA: MIT Press, 2012.

Gradin, Carlos. "Un foco en la selva de los medios. Revista Hexágono (1971–1975)." *Revista laboratorio* 4 (2011), accessed April 6, 2012. www.revistalaboratorio.cl/2011/06/un-.

Hamerman, Sarah E. *Beyond Bookworks: Ulises Carrión's Cultural Strategies*. New York: Pratt Institute, 2017, 82 pp. Master's thesis.

Hellion, Martha. "Artists Books from Latin America." *Printed Matter: Critical Essays*, November 2006. www.printedmatter.org/researchroom/essays/latin.cfm

Hellion, Martha. *Ulises Carrión: Personal Worlds or Cultural Strategies?* Mexico City: Turner, 2002.

Hendricks, John, ed. *Fluxus Codex*. New York: Harry N. Abrams, Inc., in association with the Gilbert and Lila Silverman Fluxus Collection, Detroit, MI, 1988.

Higgins, Hannah. *Fluxus Experience*. Berkeley, CA: University of California Press, 2002.

Macias Osorno, Vania. "Schmuck y la Beau Geste Press: Todos somos uno." *Reflexiones Marginales* 6, no. 41 (October–November 2017). http://reflexionesmarginales.com/3.0/schmuck-y-la-beau-geste-press-todos-somos-uno/.

Marcin, Mauricio. *Arte Correo en un Libro*. Mexico City: Museo de la Ciudad, 2011.

Perkins, Stephen. "Artists' Periodicals and Alternative Artists' Networks: 1963–1977." PhD diss, University of Iowa, May 2003.

Perkins, Stephen, ed. *Assembling Magazines: International Networking Collaborations*. Iowa City: Plagiarist Press, 1996.

Perkins, Stephen. "Utopian Networks and Correspondence Identities." *Iowa University Website* (2006), accessed February 11, 2009. http://sdrc.lib.uiowa.edu/atca/subjugated/two_5.htm.

Perneczky, Géza. *The Magazine Network*. Translated by Tibor Szendrei. Cologne: Soft Geometry, 1991.

Saper, Craig. "Intimate Bureaucracies and Infrastructuralism: A Networked Introduction to Assemblings." *Postmodern Culture* 7, no. 3 (May 1997).

Saper, Craig. *Networked Art*. Minneapolis: University of Minnesota Press, 2001.

Schmidt, Christian. "Meanings of Fanzines in the Beginning of Punk in the GDR & FRG." *Volume* 5, no. 1 (2006): 47–72.

Schranean, Guy, ed. *Ulises Carrión, Dear Reader: Don't Read.* Madrid: Museo Nacional Centro de Arte Reina Sofía, 2015.

Smith, Owen. *Fluxus: The History of an Attitude.* San Diego: San Diego State University Press, 1998.

Tejo, Cristiana. *Paulo Bruscky: Arte em Todos os Sentidos.* Recife: Editora de Pernambuco, 2009.

Waldeck, Mila. "Perfect Unbinding: The Production and Circulation of Beau Geste Press Editions." *Journal of Artists Books* no. 41 (Spring 2017).

Walker, Ian. *City Gorged with Dreams: Surreal Documentary Photography in Interwar Paris.* Manchester: Manchester University Press, 2002.

Welch, Chuck, ed. *Eternal Network: A Mail Art Anthology.* Calgary: University of Calgary Press, 1995.

Welch, Chuck. *Networking Currents: Contemporary Mail Art Subjects and Issues.* Calgary: Sandbar Willow Press, 1986.

5 The life of others

Collecting and archiving the Cuban surveillance regime[1]

María A. Cabrera Arús

The triumph of the Cuban Revolution on January 1, 1959, ushered deep changes in the country's sartorial and material culture.[2] The economic war with the US, the implementation of a nation-wide rationing system, and the Sovietization of the Cuban economy had a profound impact not only on the material environment within the country but also on individual and collective imaginaries. However, when in 2012 I began conducting research on the politics of materiality during the first three decades of Cuban socialism, scholars were mostly oblivious of these and other transformations.[3]

I had initially planned on studying the politics of automobiles, fashion, and domestic spaces in Cuba during the 1970s and 1980s – I would eventually narrow down the topic to the politics of clothing – and needed a methodological instrument to gather as much information as possible on the meanings of objects and practices during the Cold War past. To overcome the dearth of primary data resulting from the loss or lack of access to state archives, I created a blog called Cuba Material in 2012.[4] I published images of artifacts along with all the information I could gather on them, expecting readers to comment with the meanings these objects had to them during the years I was studying. Soon, it was clear that the blog was making more of a significant impact as a methodological instrument, but was also a digital hub to archive – and share – the material culture of Cuban socialism. I decided then to complement the virtual repository with a physical collection dedicated to the private sphere and domestic space.

In his groundbreaking book *The Empire of Fashion*, French philosopher Gilles Lipovetsky links the "growing collective taste for the past" with the proliferation of memorials.[5] By the end of the century, he says, the field of patrimony and commemoration had expanded to the extent that "practically everything is subject to preservation, even things that are very modest or not at all remote (artifacts from the 1950s and 1960s)."[6] Such is the case of Cuba Material, an extensive collection dating to a period framed between 1959 and 1989. "Commemorative bulimia," Lipovetsky warns us, might translate into simplified recreations of yesteryears that obliterate the meanings things and styles actually had, a caveat always taken into account when curating and exhibiting items from the Cuba Material collection.[7] Along with the ephemera and artifacts preserved, the blog documents the meanings of associated practices – micro and macro, private and public, domestic and institutional.

This is not to celebrate Cuba's socialist past, but to document its materiality for further reference and/or research. Since 2012, for instance, it has been a source of data to scholars and journalists, apart from being a great aid in my own research on the politics of fashion under Cuban socialism, which demonstrates that clothing was a mechanism of representation and impersonal rule in the island during the Cold War past.[8] In the style, labels, and materials of the clothing and footwear produced by the government and individual consumers, I identify narratives of nationalism, egalitarianism, and modernization crafted to manufacture consent, understanding some preferences for foreign clothes and styles – such as sandals, tight jeans, and miniskirts – or refusals to wear normative clothes – such as the school children's uniform handkerchief – as practices of resistance and protest. I finally argue that fashion was fundamental to the Cuban regime to produce a revolutionary figured world based on olive-green fatigue uniforms, nationalist guayabera shirts, and work clothes, as well as to individuals to convey resistance and dissent.[9]

With almost 400 blog posts and thousands of physical objects and documents stored between the United States and Cuba, Cuba Material is probably the largest repository of material culture of Cuban socialism.[10] The collection includes appliances, furniture, clothing, ephemera, and memorabilia produced, commercialized, or used in Cuba between the revolutionary triumph in 1959 and the collapse of the state socialist regimes of Eastern Europe and the disintegration of the USSR after 1989. I obtained many items from my own family, picked some others up from city sidewalks, and bought even more at secondhand stores and on the Internet. Others are gifts and donations from friends, acquaintances, and readers of the blog.[11]

The scholar of Slavic literatures and languages Svetlana Boym, like Lipovetsky, has associated the increasing interest manifested in postsocialist societies toward the socialist materiality with a "'deideologized' attitude," which she deems oblivious of past collective traumas and only interested in satisfying the tourists' gaze.[12] In her view, postsocialist processes of memorialization promoted in the media and mainstream culture do not favor critical thinking nor contribute to the documentation of the socialist past, insofar as they are mostly produced as props to recreate former utopias and totalitarianisms.[13]

Spared from *glasnosts* and Velvet Revolutions, the Cuban state socialist regime did not collapse with its Soviet bloc allies, remaining attached to the brand "socialism" and to institutions created, in most cases, during the 1970s after the Soviet model. Taking into account that, to a great extent, these particularities continue defining Cuba's present, it should not be expected that the socialist materiality evoke meanings even remotely similar to those observed in Eastern Europe and the former Soviet republics that transitioned to neoliberal market capitalism. In Cuba, for instance, COMECON appliances and automobiles are still in use, languishing in rundown living rooms and state offices, or serve as spare parts to repair other appliances and fixtures.[14] In comparison, the material culture of the prerevolutionary years does gain traction, sold in antique shops or decorating commercial spaces.

Especially ignored by specialists and discarded by those people who once possessed them are the identity cards, certificates, diplomas, sales receipts, user manuals, and warranty notices the Cuban government and its institutions produced after 1959 as part of egalitarian, nationalist, and modernizing narratives that on occasion connected the postrevolutionary present with the prerevolutionary past or with the Soviet Bloc and the Third World imaginaries. For instance, the vinyl record *Intercosmos*, commissioned by the Cuban Ministry of Culture and the Department of Revolutionary Orientation (DOR) as part of the propagandistic campaign of the USSR–Cuba joint space flight, praises the friendship between the people of the Soviet Union and Cuba while also celebrating Cuban nationalism and its Latin American roots.[15]

But the politics of identity cards, certificates, diplomas, and other memorabilia produced by the Cuban government and its institutions after 1959 goes beyond discourses that legitimized the regime. They are all disciplinary instruments. Through them, the government exerted its power to control, regulate, standardize, and classify individuals, turning them, as Michel Foucault puts it, into "objects and [. . .] instruments of its exercise."[16] They are, in other words, the manifestation of a "modest, suspicious power" and its "minor procedures," apparently unrelated to "the majestic rituals of sovereignty or the great apparatuses of the state" yet, nevertheless, fundamental to the consolidation and stability of the system, as Foucault maintains.[17]

In the following sections, I will discuss some of the forms in which this power was manifested in Cuba, examining a series of documents and ephemera in the Cuba Material collection – a Military Registry coupon, a workers' census receipt, workers' and students' personal files, and a gamut of ID cards issued by institutions of the Cuban government – and will argue that in their function of control and classification of individuals they also doubled as mechanisms of discipline and surveillance. The printed matter examined in the following pages dates from the 1960s and early-1970s, the period of institutionalization of the postrevolutionary regime.

All people's war

After the 1959 victory, the political regime invested many symbolic resources in the reification of the revolutionary ethos in daily postrevolutionary life.[18] Some of the principal symbols that conveyed this view were the sartorial attributes and practices of the Sierra Maestra guerrilla: fatigue uniforms, beards, hiking trips to the Sierra Maestra mountains, etc. These became more than characteristic traits of the new leadership, as the general public also began to incorporate them. The notion of a *pueblo uniformado*, that is, a uniformed society or a society in uniform, conveyed the idea of a country in permanent revolution, in which, as Castro suggested, "all the people [we]re soldiers of the revolution."[19] With the militarization of practices and identities, as the official website of the Cuban Revolutionary Armed Forces (FAR) states, the defense of the country "stopped being a task only assigned to the military to become a mission of the people."[20]

Putting the notion of the all people's war into practice, early in 1959 the government began to provide military training to brigades of workers, students, and school-age children, organized into voluntary militias consolidated after October 1959 under the National Revolutionary Militias.[21] By the end of 1960 this organization counted over 500,000 effectives, but only three years later it had lost momentum, riddled with issues of discipline and difficulties controlling the troops, which were fully disarmed by 1965.[22] Authorities then dedicated themselves to the development of a professional army, and militiamen turned into (reserve) soldiers.[23]

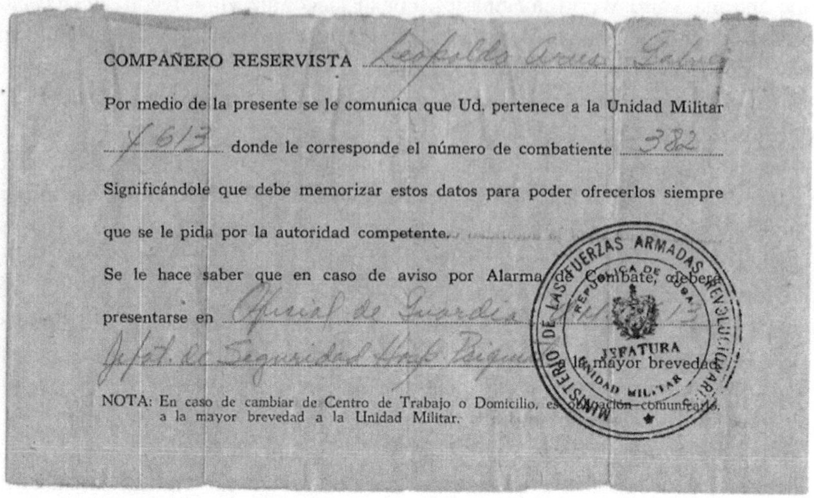
COMPAÑERO RESERVISTA ....

Por medio de la presente se le comunica que Ud. pertenece a la Unidad Militar Y 613 donde le corresponde el número de combatiente 382

Significándole que debe memorizar estos datos para poder ofrecerlos siempre que se le pida por la autoridad competente.

Se le hace saber que en caso de aviso por Alarma de Combate, deberá presentarse en ....

NOTA: En caso de cambiar de Centro de Trabajo o Domicilio, es obligación comunicarlo a la mayor brevedad a la Unidad Militar.

Figure 5.1 Notification of the reservist number and military barrack assigned to a reservist (1960s).

Source: Cuba Material collection.

In July 1963, Fidel Castro announced the forthcoming implementation of mandatory military service.[24] The draft law was passed on November 26. Law No. 1129 of the Revolutionary Government created military committees, draft commissions, enrollment offices, and a military registry through which authorities would keep records of the reserve troops.[25] All males between the ages of 16 and 44 were required to register as FAR reservists, receiving a number, an ID, and an assigned barrack (Figure 5.1) where they had to report in case of a military invasion or any kind of emergency.[26]

The bureaucratic mechanisms decreed in Law No. 1129 not only helped in the militarization of society, obliterating distinctions of region and class; they also tightened state control over the working-age male population and curtailed its freedom. For instance, to emigrate, reservists had to obtain the Military Registry's clearance. Helping to guarantee the state ownership of the monopoly over legitimate violence, the administrative mechanisms and dispositives of control of the military gave shape to a panoptical bureaucracy, which as the following sections show was based, too, on the surveillance of workers, students, and citizens in general.

A workers' state

The Cuban Revolution, a social movement of bourgeois origins, radicalized once it seized power to give place to a state socialist regime of the Soviet type that claimed to represent the interests of the working class.[27] "The revolution in Cuba," sociologist Maurice Zeitlin observes, referring to the postrevolutionary regime, "is a working-class revolution; the workers in the cities, in the sugar centrals, and in the countryside are its social base."[28] Policies that targeted and controlled the labor force – highly organized in the prerevolutionary years – became the priority of a political grammar that placed the working class at the center of the (post)revolutionary symbolic imaginary.[29] Many of these policies actually favored workers, improving through redistributive mechanisms their material conditions of life, especially among the most destitute sectors, and democratizing access to education, health care, mass consumer goods, and overall social mobility.[30] Others, however, were from their inception mechanisms of discipline and control.

On March 18, 1960, the Revolutionary Government passed Law No. 761, which called for an immediate census of the workforce (Figure 5.2) to be conducted by the Ministry of Work.[31] The following year, government agencies issued identification cards to state employees (Figure 5.3), which by the end of the decade constituted the quasi totality of the workforce after the nationalization of 57,000 private businesses during the Revolutionary Offensive of 1968 – except for a relatively small group of peasants who were allowed to work their own land.[32] Participation in the only state-controlled labor union, the Confederation of Cuban Workers (CTC), then became compulsory, and CTC-issued workers' IDs were distributed through the state payroll.[33]

Through the workers' census and ID cards, the government obtained detailed information on the workforce, including private data such as the workers' dates of birth, blood types, fingerprints, and more. These mechanisms of control and surveillance were further complemented by other CTC-issued ID cards that regulated access to the workers' clubs and to the retail stores of the rationed market.[34]

The workers' club IDs authorized union members and their immediate relatives to enter the premises of the recreational installation that the state had assigned to their union. Many of these buildings had been created in the installations of former

REPUBLICA DE CUBA
MINISTERIO DEL TRABAJO
PRIMER CENSO LABORAL

COMPROBANTE PARA EL DESEMPLEADO

DESEMPLEADO Nº 727240 -2

REPUBLICA DE CUBA
MINISTERIO DEL TRABAJO
PRIMER CENSO LABORAL

COMPROBANTE PARA EL EMPLEADO

NUMERO DEL EMPLEADO 705183-3

Figure 5.2 First Census of Workers, receipts (1960).

Source: Cuba Material collection.

Figure 5.3 Worker's identification card, issued by the Ministry of Communications (1961). Source: Gift of Mirta Suquet. Cuba Material collection.

social clubs, nationalized after 1960 and repurposed as leisure facilities for the working class. Family members were listed on the workers' IDs, along with other information such as the worker's age, home address, and union (Figure 5.4).

The impact this mechanism of control of leisure had on the development of friendship networks across class and profession, favoring interclass relationships to the detriment of networks based on class or political affiliation, is yet to be established. Arguably, it hindered the emergence of a class consciousness that would have led to the organization of, say executives, wage laborers, or professionals, to defend their interests. Moreover, this bureaucratic mechanism discriminated against unmarried and extramarital partners and children, not listed as relatives and thus de facto barred from entering the recreational facilities the government had assigned to workers' family members. Finally, some installations provided better service, consumer goods, and gastronomical offers – namely, the clubs administered by the FAR and the Ministry of Interior – and their IDs became a voucher to privileged leisure, sanctioning difference and distinction.

By the turn of the 1960s, the Ministry of Domestic Commerce (MINCIN) had divided buyers in the rationed market into groups, assigning to each a Thursday to Wednesday buying period and reserving the first day for workers.[35] The CTC-CI card provided access to stores on these days serving as a gateway to a better selection of goods and to a less-crowded shopping experience, as those were the days when stocks arrived and stores stayed open until 11 p.m.[36] This commercial privilege devised to convey the regime's long-term commitment to workers, was also a means to incentivize participation and long-term permanence in the workforce,

DECLARACION DE FAMILIARES

MADRE ....

PADRE ....

ESPOSA Modesta Forteza

HIJOS Nélida Martinez Forteza

Mirta Martinez Forteza

NOTA: SOLO EL PORTADOR DE ESTE CARNET Y SUS FAMILIARES AQUI DECLARADOS TENDRAN DERECHO A DISFRUTAR DE LOS CIRCULOS SOCIALES CON LA PRESENTACION DEL MISMO.
EL PORTADOR SERA RESPONSABLE DEL USO INDEBIDO DE ESTE CARNET.
EL TRABAJADOR QUE SE DESAFILIE O DEJARE DE PAGAR SUS CUOTAS SINDICALES PERDERA EL DERECHO A DISFRUTAR DE LOS CIRCULOS SOCIALES, EN CUYO CASO DEBERA DEVOLVER ESTE CARNET A SU SECCION SINDICAL.

LOS LUNES NO FUNCIONARAN LOS CIRCULOS.

Figure 5.4 Worker's identification card for purposes of gaining access to the workers social clubs (1960s).

Source: Gift of Mirta Suquet. Cuba Material collection.

as the administration had to update the worker's status in the CTC-CI card every three months.[37]

The bureaucratic mechanisms discussed in this section were some of the most visible dispositives of control over the workforce. Some of them gave labor unions a new raison d'ètre in an epoch when, under state control, and their very existence might have seemed redundant to many.[38] All in all, they strengthened and expanded the regime's bureaucratic panopticon.

Rationing consumption

It is widely understood that a decline in production, coupled with an increase in the purchasing power resulting from redistributive policies implemented after the revolutionary victory, caused the commercial imbalance that was at the origin of a widespread black market of basic consumer goods and necessities in the country. In an effort to ameliorate the lack of supplies and curb the black market, the government passed Law No. 1015 on March 12, 1962, ruling the creation of a national system of rationing for food, extended the following year to clothing and footwear.[39] To regulate these activities, the government also created the Offices for Provision Control or National Rationing Board, widely known as OFICODAs after their Spanish acronym.[40]

For the acquisition of the basic quota the system assigned to every citizen, OFICODAs distributed two different ration booklets, valid for a year. A ration booklet per family allocated food quotas, and a ration booklet with coupons allocated

Figure 5.5 Men's ration booklet (1970s).
Source: Cuba Material collection.

individuals a basic quota of clothing, textiles, footwear, and other industrial goods in scarce supply, slightly varying according to the recipient's gender and age (Figure 5.5).

Officials presented the rationing system as an act of justice in favor of the working class, which guaranteed equal access to mass consumer goods and protected against hoarders and intermediaries.[41] On the flip side, however, as it has been acknowledged, rationing stripped consumers their agency, as they entered de facto into a paternalistic relationship with the state, becoming passive cardholders always awaiting their assigned quota to arrive at stores.[42] Even more important in the context of this discussion is that the system of rationing offered a channel for the government to obtain information on consumers, including not only the date of birth and address of individuals and families but also their shopping behavior and preferences.[43]

Democratizing surveillance

Whereas the mechanisms described previously offered the possibility to control and spy on the military, the workforce, and consumers, other bureaucratic procedures of discipline and surveillance targeted society in general. The Committees for the Defense of the Revolution (CDRs), created on September 28, 1960, organized all adults older than sixteen at the level of street blocks to keep watch

Figure 5.6 ID card issued to the members of the CDRs (circa 1961–62).
Source: Cuba Material collection.

over their neighbors and support state campaigns such as the cleaning of public areas, vaccination, and voluntary work. Members received an ID card (Figure 5.6) that, among other information, included a quote from Castro that stated: "Being a member of the Committee for the Defense [of the Revolution] means having a spirit of sacrifice, being an example to other citizens, working, watching over the counterrevolutionaries. It also means working towards recruiting others, [that is,] being engaged in activities of proselytism."

The democratization of surveillance that the creation of the CDRs represented was far from disorganized. Each CDR appointed a *responsable de vigilancia*, a person responsible for collecting information on neighbors, including their acquaintances, visitors, social activities, and property, which they then had to pass on to authorities. This person was in command of two kinds of registries: a "registry of addresses" with the name, ID number, and address of every individual living on the block, and a "registry of means," which detailed the personal property that could be of collective use in case of war or catastrophe. Tools and resources inventoried in the latter included bladed weapons, work tools, portable lighting, sewing machines, typewriters, vehicles, radio broadcasters, telephones, recorders, domestic appliances, and cisterns, among others. One of these registries, created in 1988 by a CDR member in the upscale neighborhood El Vedado, in Havana, records three picks, 10 hand saws, 44 hammers, 20 lanterns, one wheelbarrow, 10 bicycles, 25 automobiles, 95 refrigerators, one binocular, and 70 battery radio receptors. Only five families in this affluent neighborhood owned a radio-cassette recorder, which is an indicator of the tight control the state had of information and the means of reproducing it within the country.[44]

In the Cuba Material collection there is an ID, originally found among the papers of a CDR *responsable de vigilancia*, issued by the National Center of Information (CNI), an institution of the Cuban intelligence dedicated to the compilation of detailed information on individual citizens, according to the defector Juan Reinaldo Sánchez.[45] This card (Figure 5.7) is labeled as "secret"

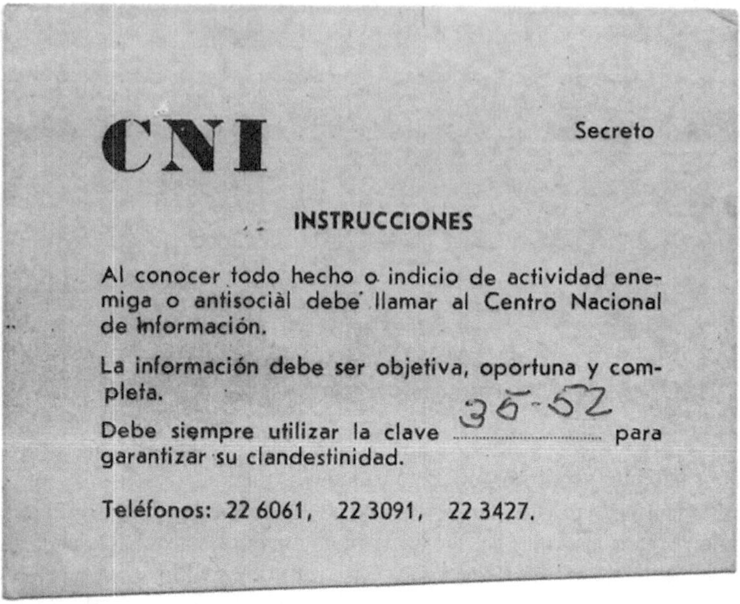

CNI

Secreto

INSTRUCCIONES

Al conocer todo hecho o indicio de actividad enemiga o antisociàl debe llamar al Centro Nacional de Información.

La información debe ser objetiva, oportuna y completa.

Debe siempre utilizar la clave 35-52 para garantizar su clandestinidad.

Teléfonos: 22 6061, 22 3091, 22 3427.

Figure 5.7 CNI card with instructions to Cuban citizens on what to spy on and how to report what they saw to the authorities, date unknown (1970s–1980s).

Source: Cuba Material collection.

and contains instructions on what kinds of suspect behaviors to look for and how to report them. To protect the confidentiality of the informant, a code was assigned that concealed his or her real name. Information on the extent, procedures, and mechanisms of recruitment of this network of espionage has proven to be hard to obtain, but it seems to have coexisted and overlapped with the CDR surveillance dispositives.[46]

The CDR members' IDs and the CNI cards document the role the Cuban state played in stimulating and sponsoring citizens' surveillance. They also expose the extent to which nonstate actors collaborated with the regime's institutions of espionage, extending and democratizing surveillance and tightening the grip the state had upon individuals and groups.

Another dispositive of citizens' control and surveillance was created on July 15 of 1971, when the government passed Law No. 1234 mandating that all citizens, including minors, obtain and carry an ID, which they had to hand over to the police and other authorities upon request. Shaped as a booklet, this document recorded the individual's full name, personal photograph, information on his or her parents, home address, and other details. Adults' IDs also recorded information such as their workplaces, professions, date of incorporation into their current employment positions, work file numbers, and criminal records. In this way, when in the mid-1970s the Cuban state socialist regime was finally fully institutionalized, the government already counted on an encompassing mechanism of control, discipline, and surveillance.

Surveilling women, peasants, and students

The bureaucratic mechanisms the Cuban regime put into place to control and surveil society also reached women, peasants, and students, groups of strategic importance within the narrative of legitimation of the fidelista state. The government also devised IDs customized for controlling and surveilling these groups, similar to the documents examined above.[47] Students, however, represent a particular case, as the government targeted them through programs of social engineering with the goal of creating the new men and women of the communist future. This task demanded even more detailed dispositives of surveillance.

In 1964, Castro told the youth:

> A day must come when we are organized in a way that everyone's personal biography is known [by the state], that [the state] keeps a personal file for each citizen, [with information] on their childhood, since the moment they enter first grade: what [the students] do, what their principal traits are, how they behave as young men, as technicians, as workers. The day must come when we have a personal file for each citizen.[48]

By the turn of the decade, the Ministry of Education had already designed a Student's Accumulative File (Figure 5.8) that allowed teachers to keep yearly record of every student's development, from kindergarten to college. The first pages collected personal information, such as the ID number, date of birth, and address, as well as information on the student's parents, including their names, dates of birth, professions, employers, and the political and mass organizations in which they participate. The following sections recorded the student's academic, disciplinary, and political merits and offenses, as well as his or her physical and psychological development, talents, study habits, and drawbacks. Serious misconducts, called "stains," were also recorded, carrying repercussions that potentially affected the student's opportunities to succeed in school. Because students never had access to their personal files, they had no way to know about their content, so they learned to police themselves to fulfill the authorities' expectations and avoid having their files stained as norm. The whole system, moreover, potentially biased teachers in favor, or against new incoming students who had problematic records, arguably further impacting their future.

Many other IDs were produced by the regime both during its first two decades in power and afterwards. Disciplinary spaces, Foucault observes, "tend to be divided into as many sections as there are bodies or elements to be distributed."[49] But the examples discussed in this chapter provide a general view of the methods and dispositives of the Cuban surveillance bureaucracy.

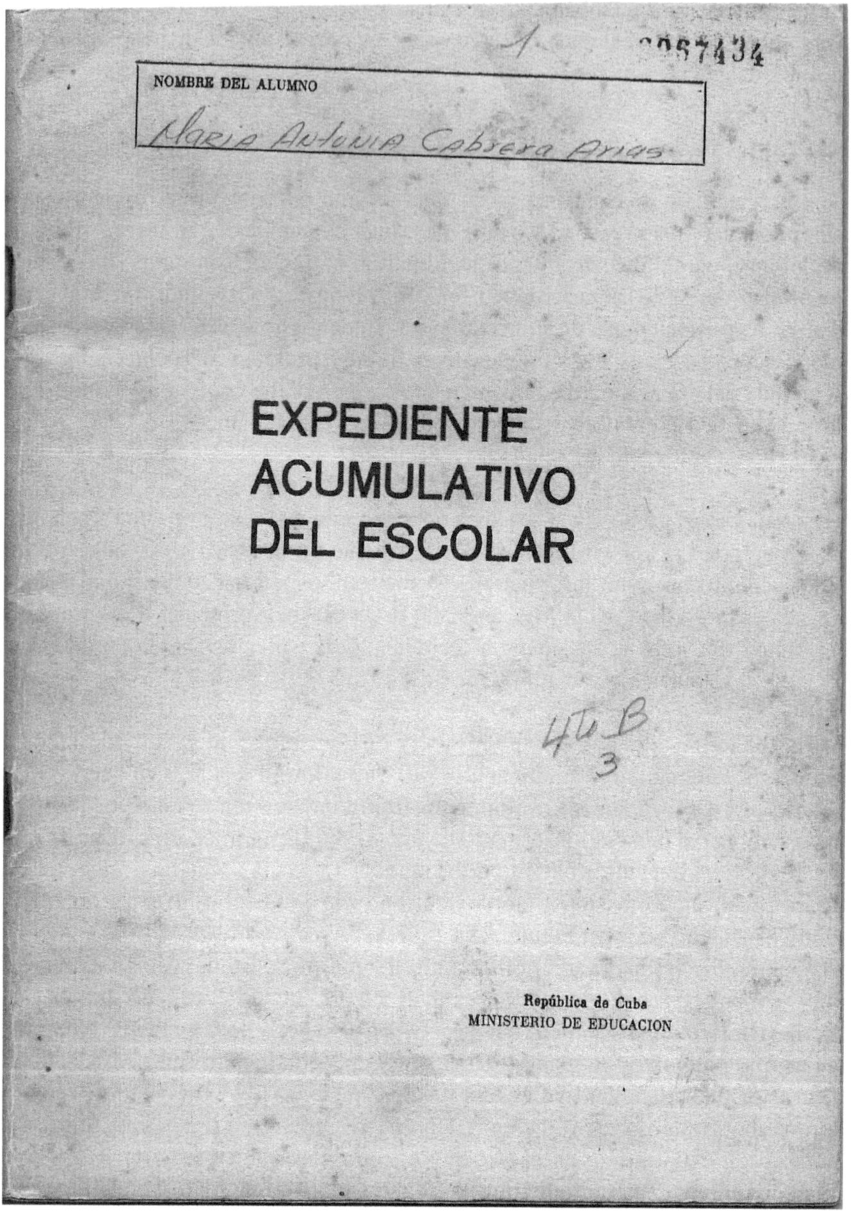

NOMBRE DEL ALUMNO

Maria Antonia Cabrera Arias

EXPEDIENTE
ACUMULATIVO
DEL ESCOLAR

República de Cuba
MINISTERIO DE EDUCACION

Figure 5.8 Student's Accumulative File (1978).

Source: Cuba Material collection.

Conclusion

All in all, the universe of ID cards and personal records the Cuban government produced after 1959 attest to the extent and scope of the regime's surveillance of individuals and groups, illustrating as well some of the mechanisms of socialization, coercion, and discipline in socialist Cuba.[50] The documents examined in this chapter constitute dispositives of surveillance, yet also demonstrate how citizens collaborated with the regime's disciplinary apparatus, either by spying on each other and passing this information on to authorities or by policing themselves to fulfill the regime's expectations.

Now, 60 years after the revolutionary victory, in Cuba many state archives are closed to public scrutiny, hindering the possibility of studying in detail the dynamics discussed. The documents from the Cuba Material collection might elicit nostalgia for members of several generations, but they attest, above all, to the impact and extent of the mechanisms of control, discipline, and surveillance the Cuban regime put into place after the 1959 revolution. This is particularly important because, as David Crowley and Susan Reid argue in the introduction to *Style and Socialism*, the socialist era "is slowly passing from memory into history [and] the material environment that it fashioned is slipping away more quickly."[51] Although these authors had in mind the societies of Eastern Europe and the former USSR, where socialism did actually begin to disappear after 1989, Cuba does not escape from this fate. Ironically, in many of the countries where the state socialist regime collapsed, the material traces of this past are memorialized and consumed, either in the form of nostalgia or souvenirs. In Cuba, however, where the state socialist regime continues to hold the reins of power, this material culture is silent and inexorably disappearing, displaced by the impetuous advent of global capitalism.

We cannot know the meanings future generations will give to the material culture from Cuba's changing present and socialist past, but we can certainly safeguard, for them and for us, the artifacts our contemporaries and prior generations used, and the meanings with which they invested them. But, like the dispositives of surveillance of the Cuban regime or any totalitarian state, we can get a glimpse, through these kind of archives and collections, of the lives of others.

Notes

1 An earlier version of this chapter was first presented at "Summoning the Archive: A Symposium on the Periodical, Printed Matter, and Digital Archiving," organized by Meghan Forbes at the Institute for Public Knowledge, New York University, May 11–13, 2017. Participation in the symposium and the writing of this article were supported in part by the 2016–2017 Mellon Sawyer postdoctoral fellowship, Cuban Futures beyond the Market, I was granted by the King Juan Carlos I of Spain Center at New York University.

2 María A. Cabrera Arús, "Thinking Politics and Fashion in 1960s Cuba: How Not to Judge a Book by Its Cover," *Theory & Society* 46 (2017): 411–428; María A. Cabrera Arús and Mirta Suquet, "La moda en la literatura cubana, 1960–1970: tejiendo y destejiendo al hombre nuevo," *Cuban Studies* (forthcoming).

3 Scattered references to some political meanings conveyed through sartorial practices are found in Lillian Guerra, *Visions of Power in Cuba: Revolution, Redemption, and*

Resistance, 1959–1971 (Chapel Hill: University of North Carolina Press, 2012); Louis Pérez, *On Becoming Cuban: Identity, Nationality, and Culture* (Chapel Hill: University of North Carolina Press, 1999).

4 María A. Cabrera Arús, Cuba Material, http://cubamaterial.com.

5 Gilles Lipovetsky, *The Empire of Fashion: Dressing Modern Democracy* (Princeton, NJ: Princeton University Press, 1994). See also, Walter Benjamin, *Illuminations: Essays and Reflections* (New York: Schocken, 1968); Jacques Derrida, *Archive Fever: A Freudian Impression* (Chicago, IL: University of Chicago Press, 1995); Daniel Miller, *The Comfort of Things* (Cambridge, MA: Polity Press, 2009); Susan Stewart, *On Longing: Narratives of the Miniature, the Gigantic, the Souvenir, the Collection* (Durham, NC: Duke University Press, 1993).

6 Lipovetsky, *The Empire of Fashion*, 248.

7 Ibid.

8 Cuba Material is featured in the *American Historical Review*'s guide to freely accessible online collections of primary sources. Items from the collection are discussed in Lillian Chase, *Heroes, Martyrs & Political Messiahs in Revolutionary Cuba, 1946–1958* (New Haven: Yale University Press, 2018); Elzbieta Sklodowska, *Invento, luego resisto: El Período Especial en Cuba como experiencia y metáfora (1990–2015)* (Santiago: Cuarto Propio, 2016); Raúl Rubio, *La Habana. Cartografías culturales* (Valencia: Aduana Vieja, 2013); Abel Sierra Madero, "'El trabajo os hará hombres.' Masculinización nacional, trabajo forzado y control social en Cuba durante los años sesenta," *Cuban Studies* 44 (2016): 309–349. They also appear in journal articles and online publications, including Nora Gámez Torres, "Cuba tendría archivos similares a los de la Stassi alemana," *elnuevoherald.com*, October 26, 2014, www.elnuevoherald.com/noticias/mundo/america-latina/cuba-es/article3390649.html; Shaday Larios, "Objetarios 1. Cuba Material: Archivo de la materialidad cubana," *Titeresante*, Julio 12, 2016, www.titeresante.es/2016/07/objetarios-1-cuba-material-archivo-de-la-materialidad-cubana-por-shaday-larios/. I develop these ideas in the manuscript *Dressed for the Party: Fashion and Politics in Socialist Cuba (1959–1989)*, in preparation based on the Ph.D. dissertation of the same title.

9 See, also, María A. Cabrera Arús, "Fashioning and Contesting the Olive-Green Imaginary in Cuban Visual Arts," in *A Movable Nation: Cuban Art and Cultural Identity*, edited by J. Duany (Gainesville: University Press of Florida, forthcoming 2019); María A. Cabrera Arús, "The Material Promise of Socialist Modernity: Fashion and Domestic Space in the 1970s," in *The Revolution from Within: Cuba, 1959–1980*, ed. M. Bustamante and J. Lambe (Durham, NC: Duke University Press, forthcoming March 2019); Cabrera Arús, "Thinking Politics and Fashion."

10 Other private archives contain materials on specific periods of Cuban socialist history, notably the Ramiro A. Fernández collection of revolutionary ephemera, which extends to the early 1960s; *El archivo de Connie*, the personal archive and blog of Anna Veltfort focusing on the late 1960s and early 1970s; and the Archivo Veigas collection of art brochures and catalogs, now part of the Ella Fontanals-Cisneros Collection. All three are mostly collections of photographs and memorabilia. See, for instance, Ramiro Fernández and Richard Blanco, *Cuba Then*, revised edition (New York: Monacelli Press, forthcoming 2018); *El archivo de Connie* (blog), http://archivodeconnie.annaillustration.com.

11 I want to especially thank my family as well as Meyken Barreto, Mirta Suquet, Yasiel Pavón, Gerardo Fernández Fe, Ricardo L. Hernández Otero, Pablo Argüelles, Sergio Valdés García, Janet Vega, César Beltrán, Walfrido Dorta, María L. Pérez, Jairo Alfonso, Raúl Aguiar, and other anonymous readers for their contributions to the collection.

12 Svetlana Boym, *The Future of Nostalgia* (New York, NY: Basic Books, 2001), 57.

13 See, also, Aleksandar Bošković, "Yugonostalgia and Yugoslav Cultural Memory: Lexicon of Yu Mythology," *Slavic Review* 72 (2013): 54–78. Several articles in the *Calvert Journal*, www.calvertjournal.com, also address post-Soviet nostalgia.

14 COMECON was the organization for economic cooperation of the Soviet bloc countries. This acronym stands for Council of Mutual Economic Assistance.
15 The Joint Space Flight USSR–Cuba was launched on September 18, 1980, taking Cuban astronaut Arnaldo Tamayo Méndez aboard the Soyuz-6 spaceship along with Soviet astronaut Yuri Romanenko, a trip organized by the Soviet-led Intercosmos program.
16 Michel Foucault, *Discipline and Punish* (New York: Vintage, 1977), 170.
17 Ibid.
18 Michelle Chase, *Revolution within the Revolution: Women and Gender Politics in Cuba, 1952–1962* (Chapel Hill, NC: University of North Carolina Press, 2015); Duanel Díaz Infante, *La revolución congelada. Dialécticas del castrismo* (Madrid, Spain: Verbum, 2014); Guerra, *Visions of Power*; Yeidy M. Rivero, *Broadcasting Modernity: Cuban Commercial Television, 1950–1960* (Durham, NC: Duke University Press, 2015).
19 Fidel Castro, *Discurso a los escolares* (Havana, Cuba: Prensa y Divulgación INRA, 1959); Manuel De Paz Sánchez, *Zona rebelde: La diplomacia española ante la revolución cubana 1957–1960* (Santa Cruz de Tenerife, Spain: Centro de la Cultura Popular Canaria, 1997); Cabrera Arús, "Thinking Politics and Fashion;" Chase, *Revolution within the Revolution*; Guerra, *Visions of Power*; Rivero, *Broadcasting Modernity*.
20 *Cubagog.cu*, March 14, 2018, www.cubagob.cu/otras_info/minfar/far/Serv_militar_reserva.htm.
21 Guerra, *Visions of Power*.
22 M. L. Vellinga, "The Military and the Dynamics of the Cuban Revolutionary Process," *Comparative Politics* 8 (1976): 245–271, 247; Guerra, *Visions of Power*.
23 Ibid.
24 Ibid.
25 *CubaDefensa.cu*, March 14, 2018, http://cubadefensa.cu; *Cubagog.cu*, March 14, 2018, www.cubagob.cu/otras_info/minfar/far/Serv_militar_reserva.htm.
26 Ibid.; Guerra, *Visions of Power*.
27 Guerra, *Visions of Power*. See, also, Velia C. Bobes, *Los laberintos de la imaginación. Repertorio simbólico, identidades y actores del cambio social en Cuba* (Mexico City: El Colegio de México, 2000); Marifeli Pérez-Stable, *The Cuban Revolution* (New York: Oxford University Press, 1993).
28 Maurice Zeitlin, *Revolutionary Politics and the Cuban Working Class* (Princeton, NJ: Princeton University Press, 1967), 277.
29 Chase, *Revolution within the Revolution*; Guerra, *Visions of Power*.
30 Mayra Espina Prieto and Lilia Núñez Moreno, "The Changing Class Structure in the Development of Socialism in Cuba," in *Transformation and Struggle: Cuba Faces the 1990s*, ed. S. Halebsky and J. M. Kirk (New York, NY: Praeger), 205–218.
31 Jesus Echerri Ferrandiz, "El sistema de justicia laboral cubano: apuntes y reflexiones," *AmbitoJuridico.com.br*, March 15, 2018, http://ambito-juridico.com.br/site/index.php/%3C?n_link=revista_artigos_leitura&artigo_id=609&revista_caderno=25.
32 Guerra, *Visions of Power*.
33 Ibid.
34 Alejandro de la Fuente, *A Nation for All: Race, Inequality, and Politics in Twentieth-Century Cuba* (Chapel Hill: University of North Carolina Press, 2001).
35 In 1969, the government introduced the Plan San Germán in Havana, separating consumers into random groups identified by letters (e.g., group A, group B, group C, group D, group E), which were assigned specific days to buy their goods. Julio C. Díaz Acosta, "Consumo y distribución normada de alimentos y otros bienes," in Cincuenta años de la economía cubana, ed. O. E. Pérez Villanueva (Havana, Cuba: Ciencias Sociales, 2010), 333–362.

See also, Cabrera Arús, "The Material Promise;" Margaret Randall, *Women in Cuba: Twenty Years Later* (New York: Smyrna, 1981).

36 Ibid.; Eloise Linger, "Combining Moral and Material Incentives in Cuba," *Behavior and Social Issues* 2 (1992): 119–136; Muriel Nazzari, "The 'Woman Question' in Cuba: An Analysis of Material Constraints and Its Solution," *Signs* 9 (1983): 246–263.
37 Cabrera Arús, "The Material Promise."
38 Guerra, *Visions of Power*.
39 Díaz Acosta, "Consumo y distribución normada."
40 Ibid.; Medea Benjamin and Joseph Collins, "Is Rationing Socialist? Cuba's Food Distribution System," *Food Policy* (November 1985): 327–336.
41 Ibid.; Díaz Acosta, "Consumo y distribución normada;" Mayra Espina, Ángel Hernández, Viviana Togores, and Rafael Hernández, "Controversia: El consumo: Economía, cultura y sociedad," *Temas* 47 (2006): 65–80; Nazzari, "The 'Woman Question'."
42 Bobes, *Los laberintos*; Haroldo Dilla, "Cuba: ¿Cuál es la democracia deseable?," in *La democracia en Cuba y el diferendo con los Estados Unidos*, ed. H. Dilla (Havana, Cuba: Centro de Estudios sobre América, 1995), 169–189; Emilio Morales, *Cuba. ¿Tránsito silencioso al capitalismo?* (Miami, FL: Alexandria Library, 2009).
43 Bobes, *Los laberintos*.
44 This figure attests as well to the limited commercialization of consumer goods during the 1970s and the consumer-oriented 1980s. See Anna C. Pertierra, *Cuba: The Struggle for Consumption* (Coconut Creek, FL: Caribbean Studies Press, 2011).
45 Gámez Torres, "Cuba tendría archivos."
46 María A: Cabrera Arús, "Carné del CNI," *Cuba Material*, August 12, 2014, https://cubamaterial.com/blog/carne-del-informante/; Gámez Torres, "Cuba tendría archivos."
47 Chase, *Revolution within the Revolution*; Guerra, *Visions of Power*.
48 Fidel Castro, "Discurso pronunciado por el Comandante Fidel Castro Ruz, Primer Secretario del Partido Unido de la Revolución Socialista y Primer Ministro del Gobierno Revolucionario, en la concentración para celebrar el IV aniversario de la integración del movimiento juvenil cubano, en la Ciudad Escolar 'Abel Santamaria', Santa Clara, el 21 de octubre de 1964," www.cuba.cu/gobierno/discursos/1964/esp/f211064e.html. I am grateful to Mirta Suquet for the reference to this speech.
49 Foucault, *Discipline and Punish*, 143.
50 The impact of the Cuban regime's surveillance is also reflected in the literary and artistic production, some of which is discussed in Rachel Price, *Planet/Cuba: Art, Culture, and the Future of the Island* (London: Verso, 2015). See, also, literary works such as Eliseo Alberto, *Informe contra mí mismo* (Barcelona, Spain: Alfaguara, 2002); Antonio J. Ponte, *Villa Marista en plata: arte, política, nuevas tecnologías* (Madrid, Spain: Colibrí, 2010); Enrique del Risco, ed., *El compañero que me atiende* (Madrid, Spain: Hypermedia, 2017).
51 David Crowley and Susan Reid, "Style and Socialism: Modernity and Material Culture in Post-War Eastern Europe," in *Style and Socialism: Modernity and Material Culture in Post-War Eastern Europe*, ed. S. Reid and D. Crowley (Oxford, England: Berg, 2000), 18.

Bibliography

Alberto, Eliseo. *Informe contra mí mismo*. Barcelona, Spain: Alfaguara, 2002.

Benjamin, Medea and Joseph Collins. "Is Rationing Socialist? Cuba's Food Distribution System." *Food Policy* (November 1985): 327–336.

Benjamin, Walter. *Illuminations: Essays and Reflections*. New York: Schocken, 1968.

Bobes, Velia C. *Los laberintos de la imaginación. Repertorio simbólico, identidades y actores del cambio social en Cuba*. Mexico City: El Colegio de México, 2000.

Bošković, Aleksandar. "Yugonostalgia and Yugoslav Cultural Memory: Lexicon of Yu Mythology." *Slavic Review* 72 (2013): 54–78.

Boym, Svetlana. *The Future of Nostalgia*. New York: Basic Books, 2001.
Cabrera Arús, María A. "Carné del CNI." *Cuba Material*, August 12, 2014. https://cubamaterial.com/blog/carne-del-informante/.
Cabrera Arús, María A. "Fashioning and Contesting the Olive-Green Imaginary in Cuban Visual Arts." *A Movable Nation: Cuban Art and Cultural Identity*, edited by J. Duany. Gainesville: University Press of Florida, forthcoming 2019.
Cabrera Arús, María A. "The Material Promise of Socialist Modernity: Fashion and Domestic Space in the 1970s." *The Revolution from Within: Cuba, 1959–1980*, edited by M. Bustamante and J. Lambe. Durham, NC: Duke University Press, forthcoming March 2019.
Cabrera Arús, María A. "Thinking Politics and Fashion in 1960s Cuba: How Not to Judge a Book by Its Cover." *Theory & Society* 46 (2017): 411–428.
Cabrera Arús, María A., and Mirta Suquet. "La moda en la literatura cubana, 1960–1970: tejiendo y destejiendo al hombre nuevo." *Cuban Studies* (forthcoming).
Castro, Fidel. *Discurso a los escolares*. Havana, Cuba: Prensa y Divulgación INRA, 1959.
Castro, Fidel. "Discurso pronunciado por el Comandante Fidel Castro Ruz, Primer Secretario del Partido Unido de la Revolución Socialista y Primer Ministro del Gobierno Revolucionario, en la concentración para celebrar el IV aniversario de la integración del movimiento juvenil cubano, en la Ciudad Escolar 'Abel Santamaria', Santa Clara, el 21 de octubre de 1964." www.cuba.cu/gobierno/discursos/1964/esp/f211064e.html.
Chase, Michelle. *Revolution within the Revolution: Women and Gender Politics in Cuba, 1952–1962*. Chapel Hill, NC: University of North Carolina Press, 2015.
Crowley, David, and Susan Reid. "Style and Socialism: Modernity and Material Culture in Post-War Eastern Europe." In *Style and Socialism: Modernity and Material Culture in Post-War Eastern Europe*, edited by Susan Reid and David Crowley, 1–24. Oxford, England: Berg, 2000.
CubaDefensa.cu, March 14, 2018. http://cubadefensa.cu.
Cubagog.cu, March 14, 2018. www.cubagob.cu/otras_info/minfar/far/Serv_militar_reserva.htm.
De la Fuente, Alejandro. *A Nation for All: Race, Inequality, and Politics in Twentieth-Century Cuba*. Chapel Hill: University of North Carolina Press, 2001.
Del Risco, Enrique, ed. *El compañero que me atiende*. Madrid, Spain: Hypermedia, 2017.
De Paz Sánchez, Manuel. *Zona rebelde: La diplomacia española ante la revolución cubana 1957–1960*. Santa Cruz de Tenerife, Spain: Centro de la Cultura Popular Canaria, 1997.
Derrida, Jacques. *Archive Fever: A Freudian Impression*. Chicago, IL: University of Chicago Press, 1995.
Díaz Acosta, Julio C. "Consumo y distribución normada de alimentos y otros bienes." In *Cincuenta años de la economía cubana*, edited by O. E. Pérez Villanueva, 333–362. Havana, Cuba: Ciencias Sociales, 2010.
Dilla, Haroldo. "Cuba: ¿Cuál es la democracia deseable?" In *La democracia en Cuba y el diferendo con los Estados Unidos*, edited by H. Dilla, 169–189. Havana, Cuba: Centro de Estudios sobre América, 1995.
Echerri Ferrandiz, Jesus. "El sistema de justicia laboral cubano: apuntes y reflexiones." *AmbitoJuridico.com.br*, March 15, 2018. http://ambito-juridico.com.br/site/index.php/%3C?n_link=revista_artigos_leitura&artigo_id=609&revista_caderno=25.
Espina Prieto, Mayra, and Lilia Núñez Moreno. "The Changing Class Structure in the Development of Socialism in Cuba." In *Transformation and Struggle: Cuba Faces the 1990s*, edited by S. Halebsky and J. M. Kirk, 205–218. New York: Praeger, 1990.

Espina, Mayra, Ángel Hernández, Viviana Togores, and Rafael Hernández. "Controversia: El consumo: Economía, cultura y sociedad." *Temas* 47 (2006): 65–80.

Fernández Ramiro, and Richard Blanco. *Cuba Then* (Revised Edition). New York: Monacelli Press, forthcoming 2018.

Foucault, Michel. *Discipline and Punish*. New York: Vintage, 1977.

Gámez Torres. Nora. "Cuba tendría archivos similares a los de la Stassi alemana." *elnuevoherald.com*, October 26, 2014. www.elnuevoherald.com/noticias/mundo/america-latina/cuba-es/article3390649.html.

Guerra, Lillian. *Heroes, Martyrs & Political Messiahs in Revolutionary Cuba, 1946–1958*. New Haven, MA: Yale University Press, 2018.

Guerra, Lillian. *Visions of Power in Cuba: Revolution, Redemption, and Resistance, 1959–1971*. Chapel Hill: University of North Carolina Press, 2012.

Larios, Shaday. "Objetarios1. Cuba Material: Archivo de la materialidad cubana," *Titeresante*, Julio 12, 2016. www.titeresante.es/2016/07/objetarios-1-cuba-material-archivo-de-la-materialidad-cubana-por-shaday-larios/.

Linger, Eloise. "Combining Moral and Material Incentives in Cuba." *Behavior and Social Issues* 2 (1992): 119–136.

Lipovetsky, Gilles. *The Empire of Fashion: Dressing Modern Democracy*. Princeton, NJ: Princeton University Press, 1994.

Miller, Daniel. *The Comfort of Things*. Cambridge: Polity Press, 2009.

Morales, Emilio. *Cuba. ¿Tránsito silencioso al capitalismo?* Miami, FL: Alexandria Library, 2009.

Nazzari, Muriel. "The 'Woman Question' in Cuba: An Analysis of Material Constraints and Its Solution." *Signs* 9 (1983): 246–263.

Pérez, Louis. *On Becoming Cuban: Identity, Nationality, and Culture*. Chapel Hill: University of North Carolina Press, 1999.

Pérez-Stable, Marifeli. *The Cuban Revolution*. New York: Oxford University Press, 1993.

Pertierra, Anna C. *Cuba: The Struggle for Consumption*. Coconut Creek, FL: Caribbean Studies Press, 2011.

Ponte, Antonio J. *Villa Marista en plata: arte, política, nuevas tecnologías*. Madrid, Spain: Colibrí, 2010.

Price, Rachel. *Planet/Cuba: Art, Culture, and the Future of the Island*. London, England: Verso, 2015.

Randall, Margaret. *Women in Cuba: Twenty Years Later*. New York: Smyrna, 1981.

Rivero, Yeidy M. *Broadcasting Modernity: Cuban Commercial Television, 1950–1960*. Durham, NC: Duke University Press, 2015.

Rubio, Raúl. *La Habana. Cartografías culturales*. Valencia, Spain: Aduana Vieja, 2013.

Sierra Madero, Abel. "'El trabajo os hará hombres'. Masculinización nacional, trabajo forzado y control social en Cuba durante los años sesenta." *Cuban Studies* 44 (2016): 309–349.

Sklodowska, Elzbieta. *Invento, luego resisto: El Período Especial en Cuba como experiencia y metáfora (1990–2015)*. Santiago, Chile: Cuarto Propio, 2016.

Stewart, Susan. *On Longing: Narratives of the Miniature, the Gigantic, the Souvenir, the Collection*. Durham, NC: Duke University Press, 1993.

Vellinga, M. L. "The Military and the Dynamics of the Cuban Revolutionary Process." *Comparative Politics* 8 (1976): 245–271.

Zeitlin, Maurice. *Revolutionary Politics and the Cuban Working Class*. Princeton, NJ: Princeton University Press, 1967.

6 Contemporary art magazines

The archive in the archive[1]

Camilla Salvaneschi

In December 2017, the last issue of the thirty-three-year-old Swiss magazine *Parkett* was published, as a double and retrospective issue, titled "Expanded exchange."[2] On the spine, a photograph contributed by Sophie Calle appears with a sign that reads, "End" (Figure 6.1).

Since its founding in 1984, *Parkett*, which has offices both in Zurich and New York, has stood alone in the art-publishing field for its sophisticated and unique approach to the practice of contemporary art and publishing. At the beginning, each issue was devoted to a single artist, who would collaborate in most of the editorial processes, selecting writers, laying out the magazine's pages devoted to their work,[3] designing the magazine's spines – which interrelate across three issues, so that the image is gradually revealed through the publication of successive issues.[4] An artist's multiple, such as a poster, an object, or a small sculpture,[5] would also be produced in a limited edition that helped to fund the magazine's enterprise, so that it did not have to rely on advertising revenue. Later on, the magazine shifted to focusing on four artists per issue. Both the magazine and the supplementary artist's editions became the subject of museum exhibitions and catalogues. At the time of writing, *Parkett* is in the process of being digitized in order to make the entire collection accessible online via the publication's website, and to continue to make visible the publication's singular approach as a time capsule of the art of the last three decades. Now, when one enters the "book" section of *Parkett*'s website, a bookshelf appears with all the magazine's issues, with the spines and their integrated design on display, so that the reader can immediately perceive the finesse and the details of the collaborations with various artists. In the editorial of the last issue, the editors remember their early slogan: "*Parkett* is for keeps,"[6] emphasizing that the books are not only "collector's items, but also belong on a bookshelf – both physical and digital – where they can easily be accessed"[7] (Figure 6.2).

Indeed, the editors most often refer to *Parkett* as a book rather than a magazine. This might be related to the format, more similar to a catalogue or artist's book, than to a magazine, which, as opposed to the catalogue, is usually associated with the idea of "unprecious formats, flimsy covers, and inexpensive paper stocks"[8] supposing that "each issue will be soon rendered obsolete by the next."[9]

Figure 6.1 Cover of *Parkett*'s issue 100/101 (December 2017) "Expanded Exchange." The work on the cover is based on an edition Marlene Dumas did for the last issue.

Source: *Parkett*.

To be sure, such criterion do not apply to *Parkett*. The unfurling of a landscape across the spine of multiple issues, for instance, offers the viewer a sort of episodic presentation, or chapters in one history. In the case of the art magazine, this history may be considered that of contemporary art.

In 1976, John A. Walker described how art magazines, "because of their periodicity, are single issues devoted to contemporary art which provide 'snapshots'

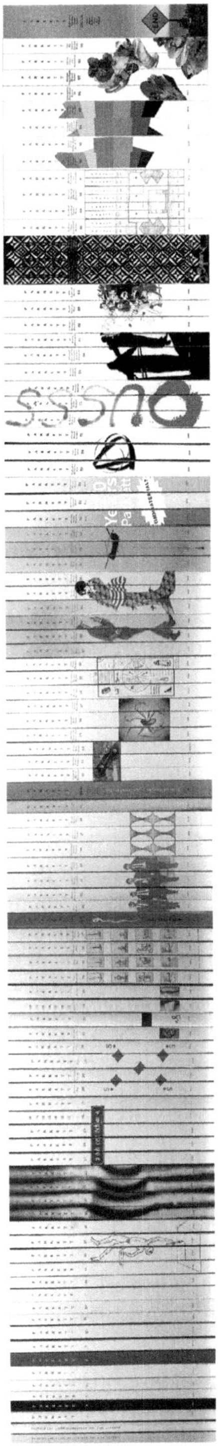

Figure 6.2 Spines of *Parkett*, issues 1–100/101.

Source: *Parkett*.

of art at particular moments. The back runs of such magazines themselves constitute a history of art, albeit an unrefined one."[10] Indeed, studying contemporary art magazines also means studying contemporary art in the very moment that it is created, defined, and legitimized. Thus, a close look at art magazines may include not only the study of art history while it is made, but also a consideration of the contemporary art system, its dynamics, and its transformation.

The decision of *Parkett*'s editors to end the publication has become "an occasion to take a clear-sighted look at the past, the present, and the future [of *Parkett*]."[11] This would seem to apply not only to the series of exhibitions they have scheduled, but for the development and construction of *Parkett*'s archive. Indeed, as will be discussed further on in this chapter, envisioning an archive is not just looking at the past and at what has been accomplished, but rather at what may be its future, what might be done with the knowledge that has been produced and how it will continue to be relevant in the future. For the construction of the archive, the editors at *Parkett* have partnered with the Suisse National Library, which is in the process of scanning all the issues. Moreover, the LUMA Foundation in Arles, for its future research center, has acquired the entire collection of *Parkett*'s editions as well as its related papers for their archive, comprising, as the editors explains, "its many photographs, original photos, designs and documents from the pre-digital era. The foundation will also take our library, which, in the course of the years, has grown to encompass several thousand volumes,"[12] including not only the issues of the magazine, but the physical library of *Parkett*'s offices and spaces, made of catalogues, artist's books, and magazines, that the editors collected over the years. The archive is conceived by the editors as a "seed bank," entailing their hope that "these seeds will have a future and will grow and inspire many other new and different projects."[13] The construction of their archive and the definition of it, marks the coming to the end of the publication, but also the envisioning of a future. The archive becomes the departing point of their legacy.

Another interesting use of the archive is that of *Cabinet*, a magazine and nonprofit organization founded in Brooklyn, New York, in 2000. In the mission statement of the magazine the editors describe the goal of the publication as a site of collection and distribution of ideas, stating:

> This is what we want *Cabinet* to be: a magazine that feels much larger inside than might have seemed possible from the outside, an alternate universe. . . . Committed to both contemporary art and to historical material, motivated by a free ranging curiosity about the world, it is an unusual art magazine, as it does not talk about artists and their work, but rather it has a first relationship to art as a work. *Cabinet* thinks of itself as reflecting the variety of interests the artists have today. A source of ideas for artists.[14]

As Graham Burnett describes further, it has been "built on the most expansive definition of 'culture' possible. This definition, influenced by contemporary anthropology, is one that implicates the observer and dismisses the idea of disinterested criticality, instead inviting engaged curiosity about the world we have made and inhabit."[15] The editor-in-chief, Sina Najafi, defined the magazine as a collection of materials that could be found on the bookshelf of an artist's library, or also as a sourcebook of ideas for artists.[16] Here again, the idea of the magazine is that of being on the bookshelf of a library, although *Cabinet*'s editors take one step further, and aim for the magazine and its monothematic character and multidisciplinary approach, to spread across the full variety of subjects on offer in a library, not merely confined to the art section.

As Sina Najafi has said in an interview, the editors have always had a deep archival respect, considering themselves compulsive keepers/collectors/conservators – a nature, that is in fact reflected in the title of the magazine itself, evoking the "*Cabinet des Curiosités*."[17] Also known in German as *Wunderkammern*, the cabinets of curiosities arose in the seventeenth century in Europe as a repository for all types of wondrous and exotic objects. These specific collections, which combined samples, drawings, and illustrations from various disciplines, from religion to science to superstition, are considered to be the precursors of the modern museum. A similar approach may be found in *Cabinet* magazine, not only for the definition of the magazine as a sort of repository, but rather for the idea expressed in the mission that an "entire alternate universe" may fit in its pages. According to Najafi, the archive becomes almost an organic continuation or external body part of the magazine, a sort of artificial intelligence, or augmented memory (Figures 6.3 and 6.4). It is essential for the very existence of the magazine itself, as a space for thinking and research, as a sourcebook, and not just for historical preservation or self-narration.

When *Parkett* announced its conclusion in February 2017 in a letter from the editors, Bice Curiger, Jacqueline Burckhardt, and Dieter von Graffenried stated that, "with the present volume of *Parkett* 99 and the following special issue 100/101 appearing this summer, the publishers have decided to bring the publication of the printed art magazine to a close. One of the major factors behind this decision is the radical change in reading behaviour brought about by our digital age."[18] Moreover, in the editorial of the last issue 100/101 the editors continue by affirming, "we still look back in astonishment at the far reaching technological advances of the past three decades and the relentlessness with which they have revolutionized the printing industry."[19] Technological improvements have allowed magazines and periodicals to flourish and proliferate all over the world, a phenomena also in evidence with the recent increase in book and magazine fairs, conferences, and exhibitions worldwide. Indeed, profound changes have taken place with digitization that have had a strong impact on reading habits. The change in readership was already lamented in 2001, during a roundtable hosted by *October* on the theme of "Obsolescence," the proceedings of which were published in the 100th issue of the journal.[20] The editors of the journal acknowledged both a discursive void in which art criticism fell to the wayside of a quickly growing market and creative industry, coupled with the lack of a coherent public, though one might argue that the lack of public lamented by *October*'s editors was not a lack, but rather a fragmentation of the public, an expansion and multiplication of local, and

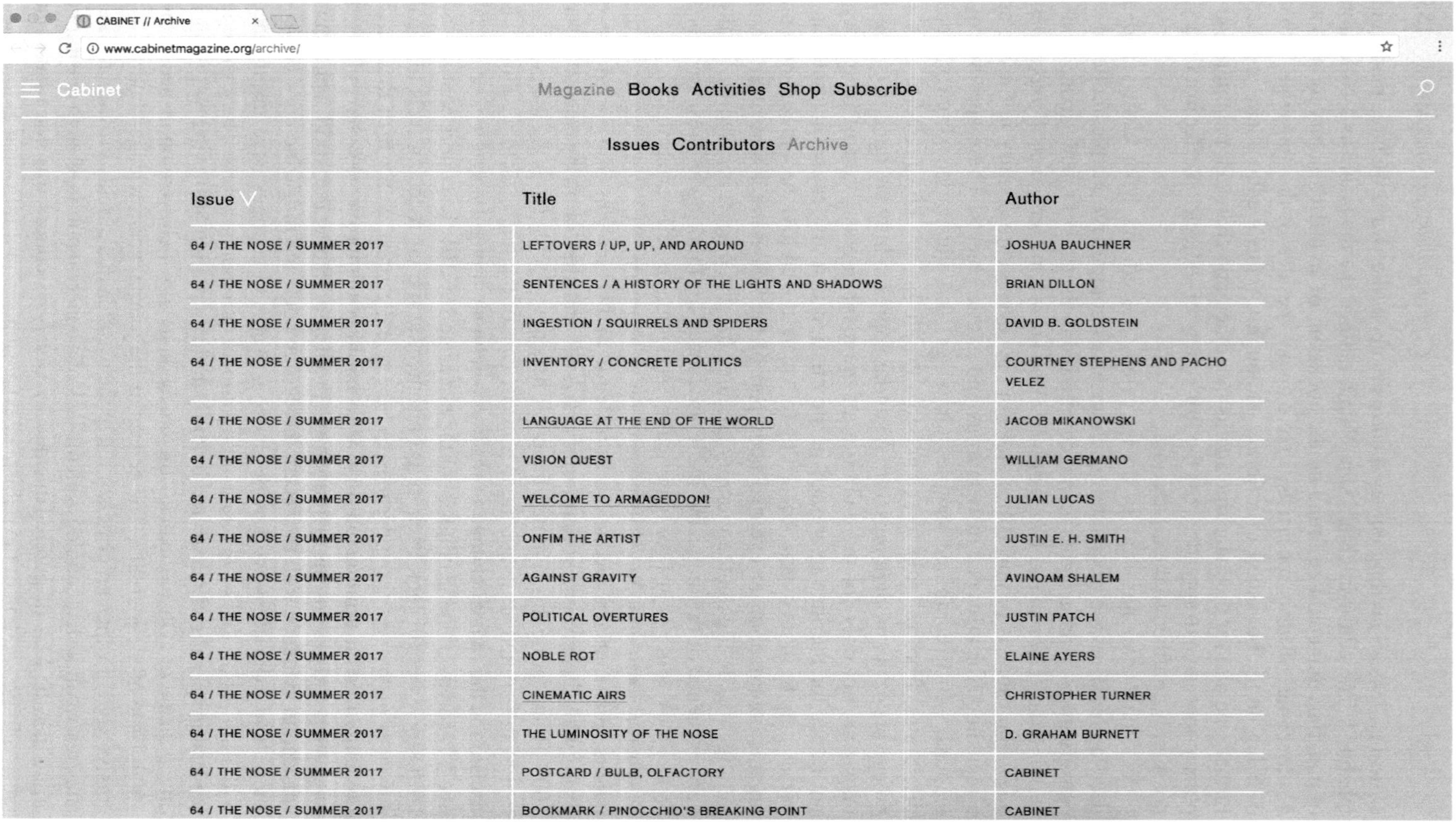

Figure 6.3 *Cabinet* magazine "Archive" page.

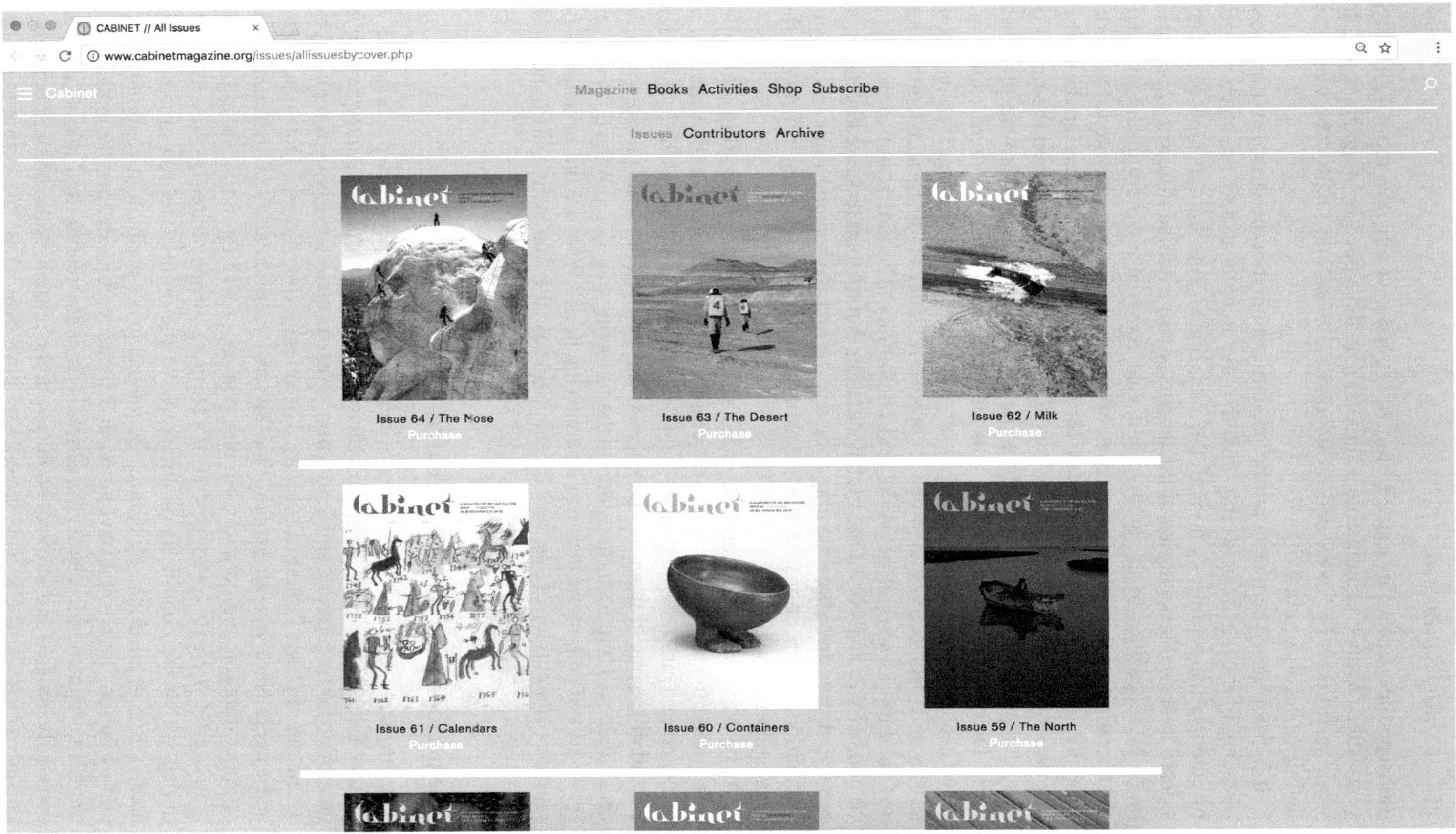

Figure 6.4 Cabinet magazine "Issues" page.

Source: *Cabinet* magazine

global publics. According to the editors of *Parkett* – writing now more than fifteen years after *October*'s roundtable – the dialogue developed by magazines and art world institutions has expanded far beyond the West, and finally "the world has definitely become round."[21]

It is from these premises that the present examination departs. I will propose a reading of the magazine deeply linked to the archive, as both are characterized by complex temporalities, and both the magazine and the archive, have been considered historical repositories of documents. I will examine different approaches made by magazines to the archive and vice-versa, in order to understand their relationship to one another. The cases offered here, diverse in editorial mission, structure, and content, all depart from the same assumption that has led the editors of *Parkett* to conclude that technological innovation has "enlarged the radius of our perception."[22] After discussing some of the magazine's archiving practices, I will look at specific magazines and their archives, in cases where editors are inventing new participatory models of publishing that challenge Western constructs and perspectives in order to pursue regional art historical legitimation and function as sites of networks and community. The aim is not to provide a comprehensive analysis of the global situation, but to point to some key cases, in order to demonstrate how the magazine and the archive become vehicles to foster international connections, and instruments to overcome the tension between the local and the global.

As both magazines and archives have such a particular relationship with time, it is crucial to study their single temporality and then to look at their relation to one another. Until now, studies have tended to focus on either on one or the other, instead of looking at them in tandem, as two aspects of the same phenomenon. In order to do so, I will focus and address the following questions: What does it mean for an art and culture magazine to be archiving itself? What are the art magazines' practices of archiving? How does the magazine use the archive as a site of network formation, to expand its own geographical boundaries?

The periodical as place, as archive

The magazine, as defined by Gwen Allen, is "a type of periodical: it is issued at regular intervals, and exists serially across a span of time."[23] Its temporality is determined by recurrence and innovation, as each consecutive issue is replaced by a new one. In "Art Periodicals Today: Historically Considered," Octavian Eşanu and Angela Harutyunyan explain that the term "periodical" – as something that implies "duration and iteration, but also pauses, or writing in time as in the German word *Zeitschrift* – is suggestive of a temporality punctuated by short intervals during which the conditions of autonomy of the modern subject are renegotiated."[24] It is implied that, by nature, the magazine is a medium in continuous flux and evolution. Its constant evolution and its periodicity defines the magazine's transient, "ephemeral" character, which is the very feature that allows it to be constantly updated and in direct contact with the present, and with our contemporaneity, which Geoff Cox and Jacob Lund have defined as "the coming together of different, but equally 'present' temporalities or 'times'".[25] Thus, the magazine's transient character, as seen in the case of *Parkett*, leads one to question the modes of historicization of the magazine itself, the strategies for the defiance of its ephemerality in order to pursue its own historical legitimation. With *Parkett*, one strategy that has been identified is the design and the transformation of the magazine into a collectable and luxurious object. Other modes of self-historicization and legitimation can be found in the magazines themselves, as they are the fundamental and primary sources of information about themselves, not only through the images and stories they edit and publish, or the evolution of the magazine from one issue to the next, but also through the ongoing conversation conducted on its pages over time.[26] A magazine might also "self-archive," with the publication of an anthologized volume as a bound and printed collection of previously published matter, or create an online platform as digital archive. *Cabinet* offers such an example with its book publication, *Curiosity and Method: Ten Years of Cabinet Magazine*, a curated selection of previously published essays – some which had been deemed successful, others defined as "instructive failures."[27]

Before entering into the specifics concerning the magazine's archive, I would like to look at the etymology of the terms archive and magazine, as for instance, the latter came to take on the meaning of "periodical" only in the first half of the eighteenth century. The term derives from the fifteenth-century French word *magasin*, meaning warehouse, depot, store, from the Italian *magazzino*, and from Arabic *makhazin*, plural of *makhzan* "storehouse," and from *khazana* "to store up" – all terms essentially referring to a storehouse, or repository of documents, an "archive." On the other hand the term archive derives from the Greek word *arkhe*̄, meaning beginning or rule, and *arkheion* which refers to a house or domicile, or in the plural, to official documents, which, according to Sven Spieker, author of *The Big Archive: Art from Bureaucracy*, refers "both to a beginning or origin and to the specific place or location where archival documents are kept."[28]

As the author of *Digital Memory and the Archive*, Wolfgang Ernst states, "in contemporary culture, for the first time, we are really living in an archive culture" and that there has never been "a culture more dynamically 'archival' than the present epoch."[29] Certainly, this archiving phenomenon may be also circumscribed within a collective anxiety to remember the amount of information and data we encounter on a daily basis. While the art magazine, as discussed above, guides us in determining what is contemporary art, the archive is the device that helps us remember that information. Thus, the archive, paradoxically, becomes the site in which so much information is held that it cannot all possibly be contained in memory, but is also a crucial instrument in ensuring that nothing will be forgotten or lost.

Derrida describes this antithetical impulse in archiving, in the aptly titled *Archive Fever*, through the Freudian principles of destruction and conservation.[30] The modernist identification of the archive envisions it as a site of accretion, in which the will to collect and to order is overtaken by its status as a space for imagination and creativity: "the archive is not representation, it is creative, and the naming of something as an archive is not the end, but the beginning of a debate."[31]

Drawing further on the Freudian conception of the mnemonic apparatus as it relates to the archive,[32] the archive might be seen as that place that allows us to "deposit" our memory in order to recover and reproduce it at any time.

On the cover of the anthology, *Curiosity and Method: Ten Years of Cabinet Magazine*, are the fox and the hedgehog. The magazine's editors, following on Claude Lévi-Strauss's notion that "animals are good to think with," state that these two creatures are "especially good for thinking." They continue:

> This is in part because they represent, in one poetic-philosophical tradition, two distinct approaches to the world around us. The curious fox, it is said, knows many small things; the cautious, methodical hedgehog, on the other hand, knows one big thing. Where the former has no appetite for systematizing, the latter is the ultimate systematizer. *Cabinet* has from its very beginning been dedicated to staging an encounter between these two outlooks in the belief that each can disrupt the familiar comforts and presumptions of the other, and that an ethics for how to understand – and therefore possibly change – the world can emerge from the friction between them.[33]

"Systematizing" and the role of the "systematizer" remain at the core of *Cabinet*'s editorial ethos, underscoring how intrinsic to its operations is the magazine's connection with the archive – the Freudian "mnemonic apparatus."

In the second half of this chapter, I extend my theoretical consideration of the interplay of the magazine and the (digital) archive, as illustrated above by *Parkett* and *Cabinet* to look comparatively at magazines that are different in approach, research, and content, but both employ the digital and physical archive to impactful ends: *Field Notes* – the online journal of the Asia Art Archive, launched in 2012 – has the aim of analyzing the rationale for an archive to publish a journal and document contemporary art in non-Western regions, and Chimurenga, a South African "platform of writing, art and politics," is an archive that assumes the format of an online library, simultaneously investigating and interrogating the role of the archive itself. Both cases see the digital and physical space of the archive more as a dynamic and active tool, and use it to promote events that may be very different in nature, but that at their core have the similar specific aim of stimulating research, networking between people and institutions, as well as facilitating the circulation of ideas on art and culture not exclusive to the West.

The magazine and the archive as method: *Field Notes* and Asia Art Archive

Since the late 1990s and early 2000s we have witnessed a proliferation, in different geographies, of archives that reflect the urge to document and preserve the history of art of specific regions outside the Western axis. One of these is the Asia Art Archive, founded in Hong Kong in 2000 with the aim of collecting materials, such as magazines, books, photographs, video and audio recordings, and documents of exhibitions, that together tell the history of art of the Asian Pacific region. The archive provides digital documentation of the history of contemporary art in the region, making it accessible worldwide, while at the same time aiming "to serve as a counter force to a growing commoditization and commercialisation of the regional art scene."[34] In June 2012, the Asia Art Archive launched a new website with the intent of providing free online access to all digitized materials. It is organized corresponding to the various types of collections represented in the archive, and is easily searchable by topic through a keyword system that annotates the materials by argument, artist, author, publication, and so forth.

It is noteworthy that alongside the 2012 relaunch of the website, Asia Art Archive also launched *Field Notes*, an online, triannual journal, published simultaneously in English and Chinese, which has the mission to provide a space for a more in-depth reflection and study of the key topics of the art and archiving fields, with particular attention on Asia. It provided a space for discussion and research, drawing from the materials of the physical archive itself. For instance, in the first issue, entitled "The And: An Expanded Questionnaire on the Contemporary,"[35] the editors address the notion of the "contemporary" in art, and how the archive might be utilized to engage and document the contemporary.

The *Field Notes* editors, Claire Hsu and Chantal Wong, take as their starting point Hal Foster's questionnaire on "The Contemporary" published in 2009 in *October* and expand upon it, both theoretically and geographically. The idea of expansion is highlighted directly in the title of the issue, and has the aim to open the conversation also to non-Western countries and thinkers (with more than forty contributors to the first issue), and understand the deep relationship between contemporaneity and the global, with a specific focus on the region. The responses to the questionnaire highlighted the need to look at the very origins and usages of the word "contemporary" as it might be different from region to region. For instance, as Karen Smith notes, in some countries "as in China, it's a lack of alternatives; 'contemporary' has become the most convenient term for art being produced in this present age . . . but 'contemporary' has little relation to a specific attitude of mind or philosophy vis-à-vis a Euro-American concept."[36] Both *Field Notes* and the work of the Asia Art Archive generally thus position themselves to engage critically with the question of the "contemporary" and its valence as an art historical term, and the editorial of this first issue closes with an invitation to the reader to actively engage with the archive in order to suggest new and future possibilities for the materials represented within the collection:

> We take this opportunity to ask that you, the reader, please take a moment to visit AAA's Collection Online. Rummage through the wealth of material from photographs to letters, testaments to ideas, conversations, events, possibilities, and crossings, contributed by artists, initiators, critics, thinkers, writers, obsessive archivists, etc. Perhaps this will recall a conversation, bound to a place, with a friend. And hopefully it can inspire a life beyond. Because it is only through working together to "reimagine", as Eileen [Legaspi-Ramirez] so gracefully puts it, that we can begin to understand what it truly means to be together, in this time.[37]

The second issue, published at the end of 2014, and titled "Archive as Method," questioned the meaning of archiving and its practices, examining the different archiving methodologies of the Asian region, be it via digital platforms, databases, maps, or networks. In this issue, *Field Notes*'s editors propose a reading of "the archive as method to illustrate the way in which initiatives like ours are taking the archive to counter, complicate, and reimagine systems in which narratives of modern and contemporary art are being produced, circulated, and understood."[38] The editors take into account John Ridener's *From Polders to Post-modernism: The Concise History of Archival Theory*,[39] allowing themselves the possibility to visualize, within the 200-year history of archival theory traced by the author, the trajectories of archival practice and to locate their position within it.[40] Ridener argues that archival theory is shaped by three main contexts: technology, paradigm, and historiography. In the introduction, Hsu and Wong state that the first of the three contexts – technology – is also their "paradigm," as in the digital everyone can be a producer and any kind of information can be easily and immediately accessed, and they thus collapse two distinct contexts, as outlined by Ridener, together. The notion of immediate accessibility, according to the *Field Notes*'s editors, brings about an urge to "rethink ownership,"[41] and to re-envision the archive. The Asia Art Archive does this by including the notion of multiplicity:

> There is multiplicity in provenance – in the case of an art archive like ours, narratives and records by the artist and/or the original archive owner are maintained. Multiple perspectives are represented during the archiving process, where the praxis of archiving is a collective and collaborative project from research to annotation. There can be multiplicity to-the-point-of excess, involving researchers, students, artists, curators, historians, thinkers, cultural workers, critics, and enthusiasts to interpret, exhibit, write, debate, connect, counter, and enrich. These pluralistic layers contribute to the redistribution of knowledge production via networks and new forms of association.[42]

The Derridian "point of excess" impulse towards collecting and archivization here becomes a generative strategy of knowledge distribution that can then be utilized and recontextualized by a wide range of actors.

These examples from the second issue of *Field Notes* suggest new, more democratic archival realities, which reflect the intentions of the Asia Art Archive itself, to challenge the conventional methods of archiving and to envision a new model of archiving that is active. In other words, the archive does not only document and collect materials that comprise the history of art in the region, but also provides better access to publications and exhibitions through their documentation, in partnership with other institutions such as universities, museums, biennials, and magazines in the region and beyond (for instance, the Asia Art Archive also has an office in Brooklyn, New York). In this way, the archive is intended to foster networks of knowledge, and promote alternative dialogues that aim to expand Western perspectives, which can be seen put into practice via the questionnaire on the contemporary cited above, or the 2018 symposium titled, "It Begins with a Story: Artists, Writers, and Periodicals in Asia."[43]

The third issue of *Field Notes* focuses on the map and on how "Asia" is defined at Asia Art Archive. The positioning of the terms "Art," "Asia," and "Archive" one next to the other, according to the editors, is already "a mapping process."[44] The map, characterized on the most fundamental level as a two-dimensional representation of a particular geographic area, would be in its very nature well suited to shifting the conception of the archive from a temporal apparatus towards a more spatial conception. The editors look at the map not just as an item, but as a tool "to trace the phenomena, practices, discourses, and developments in contemporary art,"[45] that develops across time and space, that lives and continues to unfold, in order to shape a "notion of Asia," one of the questions Asia Art Archive has been attempting to answer since its foundation. A further project, *Mapping Asia*,[46] was developed after this third issue of *Field Notes*, which is an expanded publication including works of artists, essays, reviews, and e-mail exchanges, literature, poetry, and a range of archival materials including photographs, and newspaper excerpts, cartographically manifesting a network of connections that reveals how the art of the Asian region has developed in recent decades.

The fourth and last issue of *Field Notes*, "Publics, Histories, Value: The Changing Stakes of Exhibitions,"[47] looks at the importance of the exhibition – and the curatorial, institutional, and critical system that supports it – as a primary site for the making of art history. Departing from a symposium held in 2013, "Sites of Construction: Exhibitions and the Making of Recent Art History in Asia," this final issue questions how exhibitions are remembered, and attempts to make visible some of the invisible mechanisms that lie behind exhibition histories.

Although *Field Notes* ceased publication after only four issues, the Asia Art Archive has nevertheless maintained its interest in the periodical format and the field of periodical publishing, within which there are still major gaps in research pertaining to regions outside the US and Western Europe.[48] The multiple activities and approaches that characterize the Asia Art Archive demonstrate how the archive has evolved into a "multi-functional and inter-disciplinary condition, no longer fashioned as a space only for the accrual of material, but as a platform and catalyst that enables the co-creation of meaning, where knowledge accumulation is a latent and inter-subjective process."[49] The archive in this form is no longer just a physical depository accessible to a select few researchers, but a digital tool for the public to help question and analyze the art history of a region, but also to present, with multiple approaches and projects, the materials that are part of the archive's collection in order to create new connections and networks, and inspire future projects.

The magazine's archive as site of network: the Chimurenga Library

The multiplicity of voices, and the attempt to construct the history of a specific region is also one of the aims of Chimurenga, which seeks specifically "to write Africa in the present and into the world at large."[50] Founded by Ntone Edjabe in 2002 in Cape Town, Chimurenga defines itself as a "pan African publication of writing, art and politics."[51] Among its various editorial and curatorial activities, which are all interconnected, Chimurenga sponsors events and actions (the Chimurenga Sessions), has created a radio station (PASS, or Pan-African Radio Station) and a design lab, and has launched a print publication and online library, through which Chimurenga has maintained an online and print presence. In an interview, Edjabe has stated that

> it was important for us to exist in print, in order to be able to intervene in the body of written material on and/or from Africa. I also feel the debate between online and print is an old debate. I don't think one excludes the other, and that's the approach we chose. To have a presence in both spheres.[52]

Pushing at the limits of our definition of the "periodical" is the case of *The Chimurenga Chronic* (first published in 2011), which was conceived as a single issue of the *Chimurenga Magazine*,[53] and initially not intended to be a separate periodical publication. After the first issue of the *Chronic*, which was also the last of the *Chimurenga Magazine*, the publications came to stand as two separate entities, and the *Chimurenga Chronic* was no longer published under the masthead of *Chimurenga Magazine*, but evolved into its own quarterly publication, "fostering progressive and critical discourse internationally on Africa."[54] The first issue of the *Chimurenga Chronic* (which is, again, still the last of the *Chimurenga Magazine*), is an imaginary newspaper, that, as Gwen Allen has described, "attempted to challenge the newspaper's history as an instrument of nationalism and destabilize its claims to historical truth."[55] The newspaper was backdated to the week of May 18–24, 2008, when a series of xenophobic attacks took place in South Africa. By setting the issue date in the past, *The Chimurenga Chronic* questions the way the media had reported on the events of that week, and at the same time seeks "to reactivate that history as a critical force in the present."[56] Edjabe explained that this issue of the *Chronic* was meant to function as a "time machine" that looks at the past and interrogates it.

As Edjabe wrote: "We recognized the newspaper – a popular medium that raises the perennial question of news and newness, of how we define the now and (look at) history. . . . We selected the medium both for its disposability and its longevity, its ability to fashion routine in a way that allows us to traverse, challenge and negotiate liminality in everyday life."[57] Indeed, the newspaper, as a periodical publication printed on a daily or weekly basis, is a space devoted to current events. Its engagement with contemporaneity and the now is much deeper than that of a magazine published monthly, bi-monthly, or even quarterly at a moderate price and accessible to a wider audience. The ephemerality of the *Chronic* is highlighted in the use of newsprint paper and of the tabloid format. Unlike glossy art magazines such as *Parkett*, the *Chronic* is ostensibly meant to be thrown away after use. Moreover, the editors, when choosing the newspaper format, were also interested in "circulating a disguised object" in order to reflect upon questions such as "what do we do with something valuable, qualitative, beautiful, with the density of the everyday?"[58] According to Edjabe, "the newspaper is an unexpected place to find beauty," as it is normally devoted to keeping the readers informed on current events around politics, economy, and society, with only a few pages dedicated to culture. Moreover, the newspaper can reach a much larger public, thanks to its distribution in newsstands all around the city, in train stations and airports: "a public of people we do not know, and we do not encounter."[59]

In 2009, before the launch of the *Chronic*, Chimurenga also gave birth to an online archiving project, the Chimurenga Library (Figure 6.5), which initially included information, images, and descriptions of some independent pan-African periodicals, and evolved into a space that can facilitate research projects. On the website the editors describe the library as:

> an ongoing invention into knowledge production and the archive that seeks to re-imagine the library as a laboratory for extended curiosity, new adventures, critical thinking, daydreaming, socio-political involvement, partying and random perusal. Curated by *Chimurenga*, it offers an opportunity to investigate the library and the archive as conceptual and physical spaces in which memories are preserved and history decided, and to reactivate them.[60]

The library is an online archive that indexes cultural, art, and literary magazines, both living and extinct, that have been influential for Chimurenga, but have also been important platforms of dissent in and around Africa. Each periodical is indexed with an image of the cover, a description of its the intent and mission, a list of editors and contributors under a "people" tab, a "family tree" that displays a network of similar periodicals, and bibliographic references under a "re/sources" tab. The family tree exemplifies an observation made by Didier Schulmann – librarian at the Bibliothèque Kandinsky in Paris – that a crucial factor when studying the history of a periodical publication is to consider if that magazine has been created in reference, in reaction, in response, in dialogue, or in struggle with another journal, or else, to fill a gap.[61]

The methodology behind the curation of the Chimurenga Library is described by Edjabe as "often closer to detective work, replete with entirely unexpected fortuitous coincidences, even encounters with ghosts, allegorical and otherwise,"[62] as it collects print publications that are known and admired by the editors.[63] Edjabe describes the library as a tool "to map the universe. All of the publications indexed have remained partially invisible, and the library aims to shed light onto them. The Library is our way to construct a community, as it is not so much about the publications, but about the people that edit and publish them."[64] This description of the library, recalls the different approaches to archiving that were discussed in relation to *Field Notes* and the Asia Art Archive, evidencing an interconnectedness in the ways in which disparate archival and publishing projects can be seen as integral within a broader, global network for knowledge production and community building.

With regards to the library, Edjabe also maintains that "it is also about absence, and about a system of validation that makes these publications visible."[65] Indeed, as part of its methodology, the Chimurenga Library "recognizes people as knowledge and memory as the art of the stateless."[66] The Chimurenga Library provides a list of periodicals from the different regions of the African continent, including some that are out of circulation, like the South African *Staffrider*. By providing the user/reader with information about periodicals the editors feel are relevant, the

Figure 6.5 Chimurenga Library "homepage."

Source: Chimurenga

library is actively trying to fill a gap that other institutions might not have considered. The library, still in progress, shows the ferment of the continent through its publications and most importantly through the voices of the people that edit them. The library charts a network across these periodicals and between the people who create them, in an attempt to create a subjective and "choral"[67] history of the field. Most of the periodicals indexed in the library cannot be easily accessed from the rest of the world, and for this their classification and collection in the digital library allows them to remain alive, and not be forgotten.

At its core, the library upholds the community aspect and a politics of friendship. The editors have been organizing a series of exhibitions in libraries and archives worldwide to disrupt and challenge existing library classification systems and propose an "affective" navigation of a collection that could include various writings, research, and other archival materials, such as photos, video, audio recordings, and ephemera. Hosting live events, such as talks, conversations, and performances within the physical space of the Library in Cape Town, as well as in multiple locations worldwide, the Chimurenga Library aims to circulate shared ideas within a local community and at the same time create new links and connections (Figure 6.6). For example, in regard to a recent exhibit of the Chimurenga Library held at the Showroom in London, the curator Emily Pethick writes:

> Chimurenga sent us a list of around two hundred objects to be sourced, which included books, records, films and other materials for the exhibition, about half of which they had themselves already. They proposed that we borrow everything from existing collections – we were not to "buy in" any of the objects – so the process of seeking the rest became an intrinsic part of the exhibition process. Chimurenga built the exhibition plan around what we could find, so it was a flexible map, and people kept bringing things.[68]

The ongoing project of the Chimurenga Library thus becomes an expanded, multi-sited library, that aims to challenge the potentialities of the archive, as well as that of the periodical, to stretch boundaries, create new networks of people and knowledge, and to build communities. In Emily Pethick's words the Chimurenga Library is thus a "dynamic entity."[69] The fact of naming it a "library" is no doubt to highlight the community aspect of the institution of the public library – where, in comparison to most archives, the public have more ready access to a given collection – and a wider impact and engagement from the community.

The practices discussed throughout this chapter aim to highlight the way some editorial groups and organizations are reimagining the magazine's archival role not only as an instrument of legitimation, but also as an active and participatory site of network and community, that challenges the dominant modes of mainstream publishing and archiving practices. The cases discussed do not provide a comprehensive survey of archives as developed by magazines, but present some of the very diverse processes of archival accumulation for historical purposes, and also show their potential to connect and communicate, to disseminate and socialize. The case studies of the Asia Art Archive and Chimurenga represent

Figure 6.6 Chimurenga Library, installation of Cape Town Central Library, Cape Town (2009).

Source: Chimurenga.

efforts towards making knowledge within a specific temporal or geographic scope accessible and easily researchable, and to disrupt the production and distribution of knowledge away from a Western axis. As this chapter suggests, many contemporary publishing platforms are actively thinking about the archival capacity of the publication, and evidencing a strong comprehension of the possibilities introduced by the digital, and see the self-generation of an online archive as an active tool for research and for re-envisioning histories.

To conclude with the words of the editors of *Field Notes*:

> We don't have clear answers. But for now, we do believe in the archive: in its potential to re-envision and revise the way narratives and histories are told, in its potential to reactivate, to dilute the dominant, to manifest the multiple, to challenge the meaning of ownership, and to re-define how and by whom knowledge is produced.[70]

Notes

1 The topics and ideas discussed in this paper were first presented at the "Summoning the Archive: A Symposium on the Periodical, Printed Matter, and Digital Archiving," held at the Institute for Public Knowledge (NYU) on the 12th and 13th of May, 2017. The paper has been further developed to include other case studies. It is a snapshot of the field at a particular moment: as both magazines and archives are in a state of constant flux, some of the data presented in this examination will no doubt also transform over time.

2 See Bice Curiger, Jacqueline Burckhardt, and Dieter von Graffenried (eds.), "Expanded Exchange," *Parkett* no. 100/101 (December 2017).

3 For instance, the idea of the embroidered logo came from Italian painter Enzo Cucchi. See Bice Curiger, Jacqueline Burckhardt, and Dieter von Graffenried, "Editorial," *Parkett* no. 100/101 (2017): 9.

4 The artists began designing the magazine's spines holistically beginning with issue 15. It is possible to view a complete list of the artists who designed the spines at www.parkettart.com/artists_spines. For a complete image of the whole collection of books and their spines, see also, www.parkettart.com/library-by-spines/. This image is also reproduced here in Figure 6.2.

5 For a more comprehensive idea of the editions, see www.parkettart.com/editions, accessed February 2018.

6 "Editorial," 21.

7 Ibid.

8 Gwen Allen, *Artists Magazines: An Alternative Space for Art* (Cambridge, MA and London: MIT Press, 2011), 1.

9 Ibid.

10 John A. Walker, "Art Periodicals since 1945," in *The Art Press: Two Centuries of Art Magazines*, Trevor Fawcett and Clive Phillpot (London: Art Book, 1976), 45.

11 Bice Curiger, Jacqueline Burckhardt, and Dieter von Graffenried, "Dear Reader," *Parkett* no. 99 (January 2017): 5.

12 Bice Curiger, Jacqueline Burckhardt, and Dieter von Graffenried, "Parkett at the Luma Foundation," *Parkett* no. 100/101 (2017): 40.

13 Conversation between the author and Jacqueline Burckhardt, April 26, 2017.

14 Sina Nafafi, "Editorial Statement," *Cabinet* no. 1 (Winter 2000/01): 3.

15 Sina Najafi, "Cabinet," in *Critical Machines: Exhibition and Conference*, Octavian Eşanu (Beirut: American University Beirut, 2014), Exhibition catalogue, 102.

16 Nafafi, "Editorial Statement," 3.

17 Sina Najafi, Interview, April 6, 2017.

18 "Letter to Our Readers," accessed December 2017, www.parkettart.com/media/homepage/Letter_to_Our_Readers.pdf.

19 "Editorial," 9.

20 George Baker, Rosalind Krauss, Benjamin Buchloh, Andrea Fraser, David Joselit, James Meyer, Robert Storr, Hal Foster, John Miller, and Helen Molesworth, "Round Table: The Present Conditions of Art Criticism," *October*, Obsolescence: A Special Issue no. 100 (Spring 2002): 200–228.

21 Interview with Jacqueline Burckhardt, April 2017. During an interview with the editor, Jacqueline Burckhardt said that when they started "the world was a little flatter, so they worked between the US and Europe, but since 1989 with the fall of the Berlin Wall, the world has become increasingly round. There are so many things happening in places where we have never even been. We cannot represent the art that we personally are not enough familiar with even if we wanted to, as we would need to be everywhere, to know the languages, the culture and the mentality of the place we would like to write about and to properly connect with the collaborating artists."

22 "Editorial," 21.

23 Gwen Allen, "Introduction," in *The Magazine* (London and Cambridge, MA: Whitechapel Gallery and MIT Press, 2016), 12.

24 Octavian Eşanu and Angela Harutyunyan, "Introduction: Art Periodicals Today, Historically Considered," *ARTMargins* 5, no. 3 (October 2016): 8.

25 Geoff Cox and Jacob Lund, *The Contemporary Condition: Introductory Thoughts on Contemporaneity and Contemporary Art* (Berlin: Stenberg Press, 2017), 17.

26 Jürgen Habermas, *The Structural Transformation of the Public Sphere: An Inquiry into a Category of Bourgeois Society*, trans. Thomas Burger and Frederick Lawrence (Cambridge, MA: MIT Press, 1989), 41.

27 Sina Najafi, Interview, April 6, 2017.

28 Sven Spieker, *The Big Archive: Art from Bureaucracy* (Cambridge, MA: MIT Press, 2017), 4.

29 Wolfgang Ernst, *The Delayed Present: Media-Induced Tempor(e)alities & Techno-Traumatic Irritations of "the Contemporary"* (Berlin: Stenberg Press, 2017), 9.

30 Jacques Derrida, *Archive Fever: A Freudian Impression*, trans. Eric Prenowitz (Chicago: University of Chicago Press, 1996), see in particular 12–19.

31 See the "Ten Thesis on the Archive," published on the online archive Pad.ma, accessed January 2018, https://pad.ma/documents/OH. Pad.ma "short for Public Access Digital Media Archive – is an online archive of densely text-annotated video material, primarily footage and not finished films. The entire collection is searchable and viewable online, and is free to download for non-commercial use. [. . .] The Pad.ma project was initiated by a group consisting of CAMP, from Mumbai, 0x2620 from Berlin, and the Alternative Law Forum from Bangalore. Two other organisations from Mumbai, Majlis and Point of View were part of its initiation." For further information about the project see "About Pad.ma," accessed April 2018, https://pad.ma/about.

32 Freud writes: "If I distrust my memory – neurotics, as we know, do so to a remarkable extent, but normal people have every reason for doing so as well – I am able to supplement and guarantee its working by making a note in writing. In that case the surface upon which this note is preserved, the pocket book or sheet of paper, is as it were a materialized portion of my mnemonic apparatus, which I otherwise carry about with me invisible. I have only to bear in mind the place where this 'memory' has been deposited and I can then 'reproduce' it at any time I like, with the certainty that it will have remained unaltered and so have escaped the possible distortions to which I might have been subjected in my actual memory." Sigmund Freud, "A Note Upon the Mystic Writing-Pad," in *The Standard Edition of the Complete Psychological Works of Sigmund Freud*, vol. XIX (1923–1925), *The Ego and the Id and Other Works*, ed. James Strachey, Anna Freud, Alix Strachey, and Alan Tyson (London: Vintage, 2001). Here cited from Charles Merewether, *The Archive* (Cambridge, MA and London: MIT Press and Whitechapel, 2006), 20–24.

33 Sina Najafi, "Foreword," in *Curiosity and Method: Ten Years of Cabinet Magazine* (New York: Cabinet Books, 2012), 8.

34 "Editorial Note," *Field Notes* no. 1 (June 2012): 4, accessed March 2018, https://issuu.com/asiaartarchivehk/docs/fn01_the_and_eng.

35 Sohyun Anh, Michelle Antoinette, Rasheed Araeen, Kurt Chan, David Clarke, Patrick Flores, Gao Shiming, Tapati Guha Thakurta, Atreyee Gupta, Hu Fang, Joan Kee, Naiza Khan, Hyunjin Kim, Martina K.ppel-Yang, Lee Weng Choy, Eileen Legaspi-Ramirez, Iola Lenzi, Vidya Shivadas, Karen Smith, David Teh, Reiko Tomii, and Caroline Turner, "An Expanded Questionnaire on the Contemporary," in "The And: An Expanded Questionnaire on the Contemporary," *Field Notes* no. 1 (June 2012): 11–90.

36 "Editorial Note," *Field Notes* no. 1 (June 2012): 6.

37 Claire Hsu and Chantal Wong, "Note from the Editors," *Field Notes* no. 1 (June 2012): 9.

38 "Note from the Editors," *Field Notes* no. 2 (December 2012): 4, accessed March 2018, https://issuu.com/asiaartarchivehk/docs/final_single_eng.

39 John Ridener, *From Polders to Postmodernism: A Concise History of Archival Theory* (Duluth, MN: Litwin Books, 2009), see in particular 7–20.

40 "Note from the Editors," 6.

41 Ibid., 7.

42 Ibid.

43 In January 2018, the Asia Art Archive organized, in collaboration with the University of Hong Kong, a symposium titled "It Begins with a Story: Artists, Writers, and Periodicals in Asia," to explore the role of periodicals in fostering conversation around art in twentieth- and twenty-first-century Asia. The periodical was discussed in this context not just as a container for information about art and criticism, but as an instrument for the legitimation process of emerging artists, as well as serving as a platform to create diverse and expanded publics for contemporary art and culture. The symposium was preceded by a series of exhibitions that looked at Asian art periodicals as tools for art,

criticism, and research, as well as travel writing and fiction. Another contributor to this volume, Naomi Kuromiya, was a participant in this symposium.

44 "Note from the Editors," *Field Notes* no. 3 (December 2013): 4, accessed March 2018, https://issuu.com/asiaartarchivehk/docs/131219_single_pg_sprd_final.

45 Ibid.

46 See Claire Hsu and Chantal Wong (eds.), *Mapping Asia* (Hong Kong: Asia Art Archive, 2014). The full publication may be found here, https://issuu.com/asiaartarchivehk/docs/mapping-asia-book, accessed March 2018.

47 See "Note from the Editors," *Field Notes* no. 4 (April 2015), accessed April 2018, https://cdn.aaa.org.hk/_source/ideas_documents/field-notes-04-publics-histories-value-the-changing-stakes-of-exhibitions.pdf.

48 See, for instance, Sheila M. Webb, "Magazines and the Visual Arts: The Ideal Showcase," in *The Routledge Handbook of Magazine Research: The Future of the Magazine Form*, ed. David Abrahamson and Marcia Prior-Miller (New York: Routledge and Taylor and Francis Group, 2017).

49 "Note from the Editors," 9.

50 "The Chimurenga Chronic: A Future-Forward, Pan African Newspaper," accessed January 2018, http://chimurengachronic.co.za/.

51 "About Us," *Chimurenga*, accessed January 2018, www.chimurenga.co.za/about-us.

52 Ntone Edjabe, "*Chimurenga*: Who No Know Go Know," Interviewed by Dídac P. Lagarriga with Ntone Edjabe. *Oozebap*, accessed January 2018, www.oozebap.org/text/chimurenga.htm.

53 Also the *Chimurenga Magazine* was not initially conceived as a periodical publication. In fact, it was born as a print publication, a volume with a collection of writing on the relationship between music and politics in the African context. According to Edjabe, the idea of continuing the publication came quite naturally, as after the first volume, the editors were contacted by friends and readers asking information about the publication of the next issue. [Ntone Edjabe, Interview, December 22, 2017.]

54 Gwen Allen, "Art Periodicals and Contemporary Art Worlds, Part 2: Critical Publicity in a Global Context," *Art Margins Online*, October 22, 2016, accessed January 2018, www.artmargins.com/index.php/featured-articles-sp-829273831/784-art-periodicals-and-contemporary-art-worlds-part-2.

55 Ibid.

56 Ibid.

57 Ntone Edjabe, "Letter to Readers and Collaborators of *Chimurenga Chronic*," January 2013; Allen, *The Magazine*, 224–225.

58 Conversation between the author and *Chimurenga*'s editor-in-chief Ntone Edjabe, December 22, 2017.

59 Ibid.

60 "The Chimurenga Library: About," *Chimurenga*, accessed December 2017, http://chimurengalibrary.co.za/about.

61 "L'Art en Train de se Faire," interview with Didier Schulmann, curator at the Musée national d'art moderne, and Agnès de Bretagne, librarian at the Periodicals Section of the Bibliothèque Kandinsky, accessed December 5, 2017, http://mediation.centrepompidou.fr/education/ressources/ENS-revues/ENS-revues.html.

62 "The Chimurenga Library."

63 Ntone Edjabe, Interview, December 22, 2017.

64 Ibid.

65 Ibid.

66 "The Chimurenga Library."

67 Ntone Edjabe, Interview, December 22, 2017.

68 See Kodwo Eshun, Avery F. Gordon, and Emily Pethick, "Navigating Pan-Africanisms: On the Chimurenga Library," *Afterall* no. 43 (Spring/Summer 2017): 81.

69 Ibid.
70 "Note from the Editors," 10.

Bibliography

Abrahamson, David, and Marcia Prior-Miller, eds. *The Routledge Handbook of Magazine Research: The Future of the Magazine Form*. New York: Routledge, Taylor and Francis Group, 2017.

Allen, Gwen. *Artists Magazines: An Alternative Space for Art*. Cambridge, MA and London: MIT Press, 2011.

Allen, Gwen. "Art Periodicals and Contemporary Art Worlds, Part 2: Critical Publicity in a Global Context." *Art Margins Online* (October 22, 2016). www.artmargins.com/index.php/featured-articles-sp-829273831/784-art-periodicals and-contemporary-art-worlds-part-2.

Allen, Gwen. *The Magazine*. London and Cambridge, MA: Whitechapel Gallery and MIT Press, 2016.

Alloway, Lawrence. *Network: Art and the Complex Present*. Ann Arbor, MI: UMI Research Press, 1984.

Baker, George, et al. "Round Table: The Present Conditions of Art Criticism." *Obsolescence: A Special Issue* no. 100 (October, Spring 2002): 200–228.

Benjamin, Walter. "The Author as Producer." In *Understanding Brecht*, 85–103. London and New York: Verso Books, 1998.

"The Chimurenga Chronic: A Future-Forward, Pan African Newspaper." http://chimurengachronic.co.za/

Cox, Geoff, and Jacob Lund. *The Contemporary Condition: Introductory Thoughts on Contemporaneity and Contemporary Art*. Berlin: Stenberg press, 2017.

Derrida, Jacques. *Archive Fever: A Freudian Impression*. Translated by Eric Prenowitz. Chicago: University of Chicago Press, 1996.

"Editorial." *Parkett* no. 100/101 (2017): 6–40.

"Editorial Note." *Field Notes* no. 1 (June 2012): 3–10. https://issuu.com/asiaartarchivehk/docs/fn01_the_and_eng.

"Editorial Statement." *Cabinet* no. 1 (Winter 2000/2001): 3.

Edjabe, Ntone. "*Chimurenga*: Who No Know Go Know." Interviewed by Dídac P. Lagarriga with Ntone Edjabe. *Oozebap*. www.oozebap.org/text/chimurenga.htm.

Ernst, Wolfgang. *Chronopoetics: The Temporal Being and Operativity of Technological Media*. London and New York: Rowman & Littlefield International, 2016.

Ernst, Wolfgang. *The Delayed Present: Media-Induced Tempor(e)alities & Techno-Traumatic Irritations of 'the Contemporary'*. Berlin: Stenberg Press, 2017.

Ernst, Wolfgang. *Digital Memory and the Archive*. Edited by Jussi Parikka. Minneapolis: University of Minnesota Press, 2013.

Ernst, Wolfgang. *Stirrings in the Archives: Order from Disorder*. Translated by Adam Siegel. Lanham: Rowman & Littlefield, 2015.

Eşanu, Octavian, and Angela Harutyunyan. "Introduction: Art Periodicals Today, Historically Considered." *Art Margins* 5, no. 3 (October 2016).

Eshun, Kodwo, Avery F. Gordon, and Emily Pethick. "Navigating Pan-Africanisms: On the Chimurenga Library." *Afterall* no. 43 (Spring/Summer 2017).

Fawcett, Trevor, and Clive Phillpot. *The Art Press: Two Centuries of Art Magazines*. London: Art Book, 1976.

Foucault, Michel. "The Historical a Priori and the Archive." In *The Archaeology of Knowledge and the Discourse on Language*, translated by A. M. Sheridan Smith, 126–131. New York: Pantheon Books, 1969.

Freud, Sigmund. "A Note upon the Mystic Writing-Pad." In *The Standard Edition of the Complete Psychological Works of Sigmund Freud*, vol. XIX (1923–1925), *The Ego and the Id and Other Works*, edited by James Strachey, Anna Freud, Alix Strachey, and Alan Tyson. London: Vintage, 2001. Here cited from Charles Merewether. *The Archive*. Cambridge, MA and London: MIT Press and Whitechapel, 2006.

Habermas, Jürgen. *The Structural Transformation of the Public Sphere: An Inquiry into a Category of Bourgeois Society*. Translated by Thomas Burger with Frederick Lawrence. Cambridge, MA: MIT Press, 1989.

Merewether, Charles. *The Archive*. London and Cambridge, MA: Whitechapel Gallery and MIT Press, 2006.

Najafi, Sina. *Curiosity and Method: Ten Years of Cabinet Magazine*. New York: Cabinet Books, 2012.

"Note from the Editors." *Field Notes* no. 2 (December 2012): 3–10. https://issuu.com/asiaartarchivehk/docs/final_single_eng.

"Note from the Editors." *Field Notes* no. 3 (December 2013): 3–12. https://issuu.com/asiaartarchivehk/docs/131219_single_pg_sprd_final.

"Note from the Editors." *Field Notes* no. 4 (April 2015): 4–8. https://cdn.aaa.org.hk/_source/ideas_documents/field-notes-04-publics-histories-value-the-changing-stakes-of-exhibitions.pdf.

Ridener, John. *From Polders to Postmodernism: A Concise History of Archival Theory*. Duluth, MN: Litwin Books, 2009.

Spieker, Sven. *The Big Archive: Art from Bureaucracy*. Cambridge, MA: MIT Press, 2017.

Part III

Text

7 A journal of science without boundaries

Ali Suavi's *Ulûm Gazetesi*[1]

Kenan Tekin

The history of the printing press and the ensuing emergence of periodical publications transformed politics and society around the world.[2] The long history of this technology and its products reveals that it provoked people, prompted new ideas, and stimulated new regulations. As is commonly known, the press enabled the production of large-scale publications that appealed to new audiences, while also being exploited by established institutions and classes. The history of the press and the resulting publications also brought with them increasing governmental supervision or censorship as their outreach and impact grew. In response to governmental restrictions and censorship, patrons and publishers sought various strategies to continue publishing and reaching their audiences. One such strategy was to publish abroad when circumstances at home were not amenable.

The Ottoman history of the press, though it has its own peculiarities,[3] follows the global general trends: an emergence and spread of publications together with increasing governmental restrictions and regulations. With the widespread use of the printing press and the rise of periodical publications in the second half of the nineteenth century, the Ottoman government became increasingly wary of the impact they had among its diverse communities. Hence, gradually the state imposed regulations on the press and publications in order to contain the spread of knowledge that it deemed hazardous to the fabric of its society. In this regard, reporting on conflicts after the Crimean war (1853–1856), those occurring in Lebanon and Syria, the Balkans, and Crete were particularly delicate as they involved various religious communities. Thus, governments, and particularly that of Mehmed Emin Âlî Pasha (d. 1871), were keen in keeping the criticisms of the state policies to a minimum by regulations and orders that were supposed to maintain public order, but also more likely to protect their own policies and careers against public outrage.[4] While many publishers, in line with the concerns of the state, indicated and limited the subject matter of their periodicals in order to receive necessary permissions for publication by excluding sensitive issues related to religion and politics, others had sought alternative venues to continue their work.[5] Among the latter were a number of Young Ottomans, such as Ali Suavi who in 1867 escaped to Europe. There, they found relatively safer havens for publishing.[6] Hence, they established new periodicals, beginning with *Le Mukhbir* (1867), which was followed by *Hürriyet* (1868–1870), *Ulûm Gazetesi* (1869–1870), and *İnkılâb* (1870), and included biting criticism of the government and called for its reform.[7]

This chapter analyzes *Ulûm Gazetesi* [*Journal of Sciences*], as an example of the possibilities and fragilities that came with publishing periodicals abroad. I begin with a brief description of this Paris-based Ottoman Turkish periodical, and then introduce the publisher-author, Ali Suavi (1839–1878). Subsequently I look at the content of the *Ulûm* more closely, arguing that it was a project that was rather frequently modified as Suavi single-handedly authored and published the journal, which itself engaged multiple audiences and conversations at once. Produced in Europe by an Ottoman oppositional figure, *Ulûm* reflects Suavi's multiple struggles against the Ottoman government, Orientalist representations, and overly Westernized reform projects of other Young Ottomans. To illustrate the journal's relationship with various discourses, I offer a close analysis of a few articles concerning intellectual history and reforms in the *Ulûm*.

In the literature on the period, Ali Suavi is usually represented as an unpredictable, contradictory figure.[8] Rather than merely criticizing him and his output as contradictory, however, I assert that Suavi's biography and his journal illustrate unstable patronage relationships and political changes which influence the ideas and publications of the late Ottoman intellectuals and print entrepreneurs. Furthermore, considering that the *Ulûm* was produced abroad, mostly through the efforts of a single man, it faced further challenges than publications produced in Istanbul. The obstacles that were faced were, first and foremost, financial, and a second challenge was filling the pages of the journal in time to meet the periodical's deadlines. With regards to the first, Suavi tried to overcome financial difficulty through advertisements seeking more subscriptions, as will shortly be discussed further. The second challenge was met by resorting to lists, bibliographies, calculations, serial histories, translations, and dictionary entries, in addition to a section that introduced seemingly randomly a wide range of subjects.

The Texture of *Ulûm Gazetesi*

Ulûm Gazetesi was published in Paris in 1869–1870. According to an advertisement that Suavi included in the eleventh issue of the journal, the first issue of *Ulûm* was produced in early August 1869, and the first volume, consisting of the first ten issues, was compiled on December 20, 1869. Much information about the *Ulûm* can be garnered from the various notices and advertisements in its issues. A note in the first volume described the *Ulûm* as a fortnightly journal [*risâle*]. Each issue was described as consisting of sixty-four pages, and each page consisting of seventeen lines. The price of each issue was given as 2 francs (8.5 Ottoman *kuruş*). There is an offer that those who subscribe to twenty consecutive issues at once would be given two additional issues free of charge (though in the end, only twenty-five issues in total were printed). A volume consisting of ten issues is supposed to be 640 pages long, and sold for 20 francs, which was equivalent to one French gold. In volume one, potential subscribers were directed to correspond with the following address in Paris: Suavi 44, Avenue de la Grande Armee.[9]

Apparently this was just one of the many addresses in Paris where the journal was sold. Another note on the content page of the second through the fourth issue of the *Ulûm* states that the journal is sold in Kepler Street, no. 6, in Paris. However, in notes preceding the content pages of the sixth to tenth issues, customers were directed to send requests for subscriptions to 3 rue Chateaubriand, Paris. It was not unusual for periodicals to be sold in multiple shops. However, the issues in the first volume each have only one address, suggesting that they were the main addresses for *Ulûm* at the time and that initially Suavi did not settle at a permanent address. In fact, this is confirmed by another advertisement in the eleventh issue indicating that the said issue came out late due to Suavi's moving around. However, by this point Suavi seems to have settled on the Avenue de la Grande Armee, as subsequent issues have this as the main address for correspondence. Suavi's excuse for the delay also proves that the production was not always consistent with what was advertised. To make up for any loss of subscribers, Suavi promised his customers – especially to those who made advanced payment for a one-year subscription – that he would provide them with books of equal worth for the missing issues.[10] This notion of each issue being equivalent to a book or books is compatible with Suavi's description of the issues themselves as books.

Ulûm Gazetesi consists of twenty-five issues. The first twenty issues of the journal were handwritten by Suavi himself and published with lithography, the remaining five issues were printed with moveable type (Figure 7.1). From the various announcements in the journal, beginning with possibly the third issue, we can see that Suavi was hoping to publish the journal with letterpress printing; however, he was unable to acquire the metal letters and other necessary materials for this press.[11] Letterpress printing appears to have been more expensive than lithography, since Suavi mentions that he was purposefully delaying a shift to this form of printing even though he had all the necessary equipment ready by the sixth issue – that is, roughly the end of September 1869 – so as not to have to increase the price of the next issue.[12] One reason for favoring the moveable type over lithography could be that Suavi was not a calligrapher; therefore, his handwriting, though legible, could not match the clarity of the moveable type. Second, the letterpress printing would have appeared more professional, and though Ottoman Muslims adopted it relatively late, Suavi's frequent remarks on the transition suggests that the readers also would have preferred letterpress to lithography.[13] It seems that Suavi was delaying production of images until he started using moveable type, without giving any clear reasons as to why this was not done with the lithography considering that its primary use in Europe and the Ottoman Empire was precisely to print images.

The letterpress issues of the *Ulûm* finally appeared with some significant changes. These were again noted by Suavi himself in the advertisements in the new issues. Describing the purpose of the journal [*mecmua*] as spreading the sciences and knowledge, Suavi notes that it consists of two sections: the first is an *Encyclopedic Dictionary of the Sciences and Knowledge*, arranged alphabetically;[14] the second section includes various issues and samples from the sciences. The purpose of the latter section is to narrate the sciences in the form of

١٣٦٩ علوم غزتەسی (٥١)

﴿ مسائل متفرقه ﴾

سلامان و آبسال

سلامان وآبسال قصه‌سنی ابتداء اسلامه تحدیث ایدن ذات ابن‌سینادر
اشارات نام کتابنده دیرکه « النمط الرابع فی مقامات العارفین · تنبیه ·
ان للعارفین مقامات ودرجات یخصون بها فی وهم حیوتهم الدنیا دون
غیرهم فکانهم وهم فی جلابیب من ابدانهم قدنضوها وتجردوا عنها
الی عالم القدس ولهم امور خفیة فیهم و امور ظاهرة عنهم یستنکرها
من ینکرها و یستکبرها من یعرفها ونحن نقصها علیك واذاقرع
سمعك فیمایقرعه وسرد علیك فیماتسمعه قصة لسلامان و ابسال فاعلم
ان سلامان مثل ضرب لك وان ابسالا مثل ضرب لدرجتك فی العرفان
ان کنت من اهله ثم حل الرمز ان اطقت »

امام رازی اشارات اوزرینه شرحنده دیرکه « شیخك بوراده ذکرایتدیکی
سلامان و ابسال نه‌ ایدیکی بیلمجه قبیلندن در نه‌ده قصص مشهوره‌دن ·
بونلر ایکی لفظدرکه شیخ وضع ایلدی بوکبی شیلر ایسه مستقل عقل ایله
بیلنمسی محالدر بناءعلیه شیخك حل تکلیف ایدیشی عادتا معرفت غیب
تکلیفی قبیلندندر - - - - -

بوقصه‌یی حلده چوق‌شی سویلدیلر الك اعلاسی شوقولدرکه سلاماندن
مراد آدم وابسالدن مرام جنت اولوب شیخ دیمك استرکه ادم ایله
مراد سنك نفس ناطقه‌ كدر و جنتدن مراد درجات سعادتكدر
آدمك تناول برایله جنتدن اخراجی قضیه‌سی نفسنك شهواته التفاتیله
درجات سعادتندن انحطاطی دیمکدر »

نصیرطوسی اشارات شرحنده امامه جواب صددنده دیدی که
« بن دیرم که شیخك کلامی برقصه وجودنی مشعردرکه انده سلامان

Figure 7.1 A page from the letterpress issues of the *Ulûm Gazetesi* (no 24: 1369).
Source: Courtesy of the ISAM Library.

tales so that they would be easier for the general populace to comprehend. Suavi particularly mentions the transition from lithography to moveable print, though the frequency of publication – every fifteen days – remains the same. The size of the journal changed as well with the letterpress version; the number of pages was reduced (to thirty-two pages), while the number of lines on each page was increased (to twenty-five). Suavi remarks that the journal has been enlivened with images from the sciences and crafts. Whereas Suavi had earlier suggested that the switch to letterpress printing would raise the cost of the journal, he now mentions that, beginning with the twenty-first issue, with the transition to letterpress print, the price of one issue of the journal has been reduced, to 1 franc 25 centimes, and the yearly subscription for twenty-four issues was reduced to 24 francs. In addition to the address on Avenue de la Grande-Armee, Suavi provides two alternative addresses in Paris where the journal could be bought. He claims that in the Ottoman Empire, *Ulûm* could be found in all bookstores, most likely exaggerating, as he himself being in exile, would not be able to confirm this.[15]

The letterpress issues did not last long since publication of the *Ulûm Gazetesi* was interrupted by the Franco-Prussian War that ran from July 1870 to May 1871. The last issue of *Ulûm* is dated September 1, 1870 [3 *Cemaziyelahir*, 1287]. A month later, in a notice in a temporary periodical, Suavi cited the lack of access to the printing press, paper, and cost of postal services during the war as reasons for stopping its publication.[16] After observing the situation and being convinced that peace was not foreseeable, Suavi decided to continue publishing the journal in a safe and secure place far from the Prussian military control. Hence, he moved southward to Lyon, and then from there to Marseille. While awaiting the end of the war, Suavi started publishing, again with lithography, weekly issues under the title *Muvakkaten Ulûm Gazetesi Müşterilerine* [Temporarily for Customers of the *Ulûm Gazetesi*], which were intended to inform subscribers abroad of the situation in France, bringing to their attention hot news.[17] This interim publication may also have been meant by Suavi to function as a retainer so that he did not lose his followers until he could settle somewhere more permanently and resume full operation of the *Ulûm*.

Regarding the circulation of *Ulûm Gazetesi*, we lack information on how many copies were printed or how many people subscribed to the journal in Europe and the Ottoman Empire. An interesting note on the content pages of the twenty-fourth issue states that first ten issues of *Ulûm* have been sold out, and the second volume was still available for one French gold. In a separate notice, Suavi also mentioned that he was corresponding with 300–400 people in the Ottoman Empire.[18] We might take this number as a cue that perhaps subscriptions from the Ottoman domains were around that number. In any case, it was not financially a successful enterprise as Suavi did not resume publishing the journal after the war, possibly due to lack of stable funding.[19]

Ulûm was produced mostly by the efforts of Suavi and a Turkish assistant. The print version seems to have involved some Parisian printers, since we find the following information on the letterpress issues: "Paris. – Imp. Victor Goupy,

Rue Garanciere, 5."[20] This suggests that Suavi and his assistant set the type and printed the material using facilities of a different press. In fact, while explaining the need to include a supplement to his encyclopedic dictionary of sciences and knowledge, to add missing entries, Suavi noted two mistakes that were occurring in the printing process: one of them in setting the type, and the other "while the foreign printer [*ecnebî tâbi'*] casts some type."[21] This again shows the letterpress issues required more effort and energy, and involvement of more people. Though we know little about others involved in printing the journal, we do have ample information on publisher-author Suavi, since he published his autobiography in the periodical. In addition to showing that Suavi utilized the privilege of this publishing platform to control his public image and present himself in a more sympathetic manner with the means of printing and reproduction that he held in his own hands, the autobiography helps us make a better sense of the content and intellectual influences on the making of *Ulûm*.

Ali Suavi, Young Ottomans, and the twilight of the *Ulûm*

Ulûm Gazetesi was published at a moment when Suavi and other Young Ottomans such as Mustafa Fazıl Pasha (d. 1875) and Namık Kemal (d. 1888) parted ways. During its publication, Suavi got into polemics with other lesser-known members of the Young Ottomans. It was in this context that Suavi had an urgency to write a history of the Young Ottomans, beginning with his own political autobiography.[22] We should beware that Suavi wrote this autobiography while still in exile in France, and was probably foregrounding elements that he thought would appeal to the authorities in the Ottoman Empire in providing a defense of his involvement with the Young Ottomans, or perhaps appeasing the Khedive (the viceroy) in Egypt from whom Suavi was receiving funding while publishing *Ulûm*.[23] Hence, we should be cautious about elements in this autobiography, but it is still worthy of unpacking as it reads like a statement of purpose and reveals certain information about Suavi's upbringing and political orientation that informed his writings in the *Ulûm*. Thus, in the following biography, I follow Suavi's account except where indicated otherwise.

Ali Suavi was born to a modest family in Istanbul in 1839, about a month after the well-known reform edict [*Tanzimat Fermanı*] of the same year.[24] Suavi attended Davutpaşa Rüşdiye Mektebi, equivalent to middle school, which was established according to the new system of education suggested by the commission on education [*meclis-i maârif*] as part of new reforms. His education seems to have included study of Arabic and Islamic sciences in mosques.[25] Suavi does not dwell on his early life except stating that he grew up amidst learning, and that his travels in Anatolia, Arabia, and Europe, which involved teaching and learning since the age of sixteen, were well known.[26] Suavi worked in various capacities before his escape to Europe: first as a teacher in a Rüşdiye in Bursa, later as a judge and chief secretary in Filibe (Plovdiv), where he also taught some sciences in a *medrese*.[27] Eventually, Suavi returned to Istanbul where he continued teaching and preaching in the Şehzâde Mosque.[28]

From the autobiography we learn that affected by the example of his own father and the statements of the Prophet Muhammad, Suavi grew up with an ethics of responsibility, determined to seek justice.[29] Suavi claimed that the prophetic statements inculcated in him a deep hatred of oppressors, so much so that he would not hesitate attacking the oppressor even if that meant being killed in the process.[30] This statement proved to be prophetic as it eventually came true when Suavi sought justice for the Muslim population of Filibe, and became involved in the Çırağan Palace incident in 1878, where he was murdered by the security chief of the district.

Given his fiery talks and determination for seeking justice, it appears that Suavi's preaching had made him popular enough that he was sought as the person to take charge of a new periodical. Suavi became the lead writer of this periodical, in fact a newspaper, called *Muhbir* [Reporter, 1867]. The fact that a preacher was considered to have the potential for drawing a large readership is in itself interesting, and shows how this new technology relied on the older or more conventional medium of speech and oral transmission, in order to produce print content that would draw an audience.[31] Suavi stated that his sole reason for involving himself in the newspaper business was to destroy the archaic language of the national newspapers. *Muhbir* was thus represented as the most recent and the best in a line of Ottoman newspapers, including *Takvim-i Vekâyi* (1831), *Ceride-i Havâdis* (1840), *Tercümân-ı Ahvâl* (1860), and *Tasvir-i Efkâr* (1862), which gradually broke away from the older style of writing. Suavi maintained that he destroyed the old language, and brought freedom of expression to the country.[32]

Though Suavi claimed that his goal in publishing in a newspaper was to reform the language, the following anecdote reveals that it was not the linguistic reform, but rather the political aspect of the newspaper that was most inflammatory, and caused concerns for the government. When *Muhbir* published an advertisement by Fazil Pasha on the Young Ottoman leadership (originally published in *Nord*, a Belgian newspaper), the Sublime Porte (the Ottoman government) accused Suavi of involvement in its translation from French, and publication. Notwithstanding his prior sympathy for Fazil Pasha, Suavi rejected these accusations and claimed that the owner of the newspaper had commissioned the translation without his prior knowledge.[33] Soon after this incident, Suavi published an article on the castle of Belgrade, defended the establishment of a national assembly, and was involved in arranging an aid campaign for Crete.[34] Both of these locations were going through their own political turmoil, and all of these political writings together resulted in Suavi's banishment to Kastamonu, a city in northern Anatolia, in March 1867. Curiously, it was through Fazil Pasha's invitation, financial support, and logistical arrangements that Suavi was able to escape to Paris three months later.

Fazil Pasha also arranged for the escape of two other Young Ottomans, Namık Kemal and Ziya Bey (d. 1880), to Europe. There they were supposed to print a journal in Paris, critical of the Ottoman government. It was the lure of "writing extreme pieces against the current oppressors," that encouraged Suavi and possibly others to accept Mustafa Fazil's invitation.[35] Due to the planned visit of Sultan Abdulaziz (r. 1861–1876) to the Exposition Universelle set to take place that same year (1867), Suavi and his comrades had to leave for London due to intensified French surveillance of their activities. While in London, Suavi, together with the help of other Young Ottomans, printed the first issue of *Le Mukhbir* (August 31, 1867), with the support of Fazil Pasha. This was certainly reminiscent of the preceding *Muhbir* in Istanbul, as it used the same Turkish title. *Le Mukhbir* also included a motto stating that "*Muhbir* finds a country where stating the truth is not forbidden, and comes out again."[36] This in itself illustrates the possibilities granted with relocation to new countries, which allowed opposition figures, journalists, publishers, and writers to continue their political and social messaging that was no longer possible in their home country, owing to national and international political conflicts or alliances. However, these same conditions also made it a very fragile enterprise and difficult career, as the history of Suavi's disjointed journalistic activities, his tumultuous relationship with his financial patrons, and *persona non grata* status with the Ottoman government exemplify.

Excited and determined to continue his criticism of the government, especially directed towards Âlî Pasha, Suavi soon became upset that his patron Fazil Pasha met with the Sultan during the latter's visit to Europe that year and made peace with him. Fazil Pasha was allowed to return to Istanbul where he was appointed a minister soon afterward. Suavi became critical of the Young Ottomans and Fazil Pasha. Apparently Suavi also did not produce a newspaper to the taste of his fellow Young Ottomans, and as a result of these tensions his financial support dried up. It was in this context that Suavi had to stop publication of *Le Mukhbir* with the fiftieth issue (November 3, 1868), when a worker stole press materials. Soon afterward Suavi left London, and returned to Paris, where he started publishing *Ulûm Gazetesi*.

The nature of the *Ulûm Gazetesi*

The autobiography printed in the *Ulûm* shows that Suavi grew as a Muslim political activist and became a member of Young Ottoman opposition. His early career involved contributing to or publishing periodicals that criticized contemporary Ottoman policies and called for reforms. As we have seen, escaping to Europe provided a relatively secure space to evade censorship. However, this did not last long, since the conflict among the members of the Young Ottoman Society brought an end to the publication of *Le Mukhbir*. Considering the relationship of *Le Mukhbir* to the preceding *Muhbir*, which Suavi published in Istanbul, one would expect that Suavi would continue publishing another reincarnation of these periodicals after he moved to Paris. In this light, one wonders whether *Ulûm Gazetesi* can be considered a continuation of *Muhbir* under a new guise.

Ali Suavi envisioned *Ulûm Gazetesi* as a journal that dealt with scientific and theological issues, perhaps trying to distance himself from his earlier political work, which was assumed to be a mouthpiece of the Young Ottomans. However, this is not always clear, as various descriptions of the journal present it in different terms. In the earliest advertisement, which appears before the title and content page of the third issue (dated August 16, 1869), Suavi describes *Ulûm* as a gazette that covers issues related to the sciences, adding that it will include pictures when it treats geometry, mechanics, and issues related to other sciences. Moreover, the journal was supposed to include reviews of books written by scholars [*ulema*].[37] This confirms that *Ulûm* was considered a science periodical and would be corroborated by other notices.

In an announcement on the third issue of the *Ulûm*, Suavi expresses his conviction that providing a general overview of the sciences is the first step for propagating education [*neşr-i maarif*], which he believed should be followed by bio-bibliographic information. Therefore, he promises to discuss, in each issue of the *Ulûm*, a science or any scientific issue, followed by a list of relevant books. Suavi also intended to include biographies of scholars to complement that work, with possible caveats, as he reminded the readers that being in the West he was unable to consult libraries of the East, meaning the Ottoman Empire.[38] Although Suavi provided discussions of many disciplines, including philosophy, political economy, astronomy, arithmetic, etc.,[39] biographical entries ultimately did not comprise a large portion of the publication. But this notice by Suavi indicates that he viewed *Ulûm* as a vehicle for propagating the sciences and knowledge, perhaps considering this to be a journal that would be read by students of new schools in the Ottoman Empire.

The educational orientation of *Ulûm* can be seen in a notice in the eighth issue (dated November 16, 1869). There Suavi declares that the *Ulûm Gazetesi* will include translations of one book in each science that is available in Europe. In this way, Suavi hoped to make European sciences and arts available in Turkish. In line with this intention, Suavi notes that he begins in this eighth issue with the science of trade [*fenn-i ticaret*].[40] Suavi also explains what he means by writing in a simple, descriptive Turkish, making clear that his intention is to use a simple language that consists of short and clear sentences not exceeding the most basic elements such as the subject and verb.[41] He wants to get rid of the overly long and flowery articulations that include lots of parenthetical sentences. Suavi is quite keen on this matter of language, which is a characteristic of this new medium, and is discussed in some articles in the *Ulûm*, as we will see. This attempt to simplify the language is consistent with Suavi's attempt to appeal to students and to a larger audience. In this regard, this same issue includes another change in the content of the journal, as Suavi introduces a new section under the heading of "various issues" [*mesail-i müteferrika*]. This was an effort to expand the audience further, as the section would explain matters that were incomprehensible to the layperson [*avâm*].[42]

In other parts of *Ulûm*, Suavi cast the journal in a different light. For instance, in an article on journalism, Suavi asserted that for a successful political journal or newspaper a simple language was key, money was important, and readership was crucial. Moreover, in order to obtain this readership, the journal was supposed to have an ideology. Suavi took up his own journal *Ulûm* to illustrate this last point, which according to him, stood in defense of the parliament [*millet meclisi*] and the council of the community's elite [*şuray-ı nevvâb-ı ümmet*]. According to Suavi, unless a political newspaper adheres to a specific political idea, it cannot survive.[43] Notwithstanding his previous advertisement, in this description Suavi explicitly represents *Ulûm* as a political periodical. Furthermore, Suavi rejected the idea, in the first letterpress issue – no. 21 – of reducing the journal from two sections to one so that further work on the encyclopedic dictionary could be completed sooner. Suavi stated that if he were to devote the whole journal to the encyclopedia of sciences, there would not remain any space to "condemn and curse oppressors."[44] Notwithstanding previous advertisements, this also suggests that Suavi still saw the *Ulûm* as a tool for political criticism. Based on Suavi's descriptions, we can say that *Ulûm* was mainly intended as a journal of popular science; however, it was also considered as an instrument of political change.

The content of *Ulûm Gazetesi*, in fact lends itself to Suavi's various descriptions since it included a number of issues and themes ranging from political economy, contemporary history, philosophy, philology, geology, astronomy, mathematics, matters concerning popular religious themes, and crafts such as printing. The periodical was supposed to defend theological matters; however, such issues appeared infrequently, and constituted a small percentage of the total publication. The following excerpt from the contents of the first volume of the *Ulûm* illustrates a diversity of themes in this popular science journal:

> "Turk"
> "God is the sovereign (with a register of books on politics)"
> "Analysis of the history of *Ishkanyan* [Parthians]"
> "History of the magnetic power"
> "Turkish language and alphabet"
> "Metric poetry (science of *Arûz*)"
> "Intellectual history (wisdom and philosophy)"
> "Introducing *Kashāf Istilāhāt al-Funūn* which is published in Calcutta"
> "Definition of the science of arithmetic (includes a bibliography of math books)"
> "Interpreting verses on diverging inclinations"
> "Strange Oriental Knowledge" [*Garaib-i Maarif-i Şarkiye*]
> "Amalekites, Phoenicians and the Greeks"
> "That Phoenicians were of the Arab Amalekites"
> "That Greeks originated from Arab Amalekites"
> "Calculating the Debt and tables"
> "Amourtier *mevkuf*"
> "The problem of interest and its calculation as well as its history"
> "Planets: definition and explanation"

This excerpt of the contents for the first volume alone suffices to illustrate that the *Ulûm Gazetesi* contained a range of subjects, including intellectual history, language and poetry, politics and economy, ancient history, and various mathematical and natural sciences. This diversity of themes can be also observed in the section of the journal that was devoted to strange and exotic matters for the enjoyment of the common people. Considering the audience of the journal, conscious adoption of a simple idiom, and an array of the issues that could appeal to multitudes, this journal, like others that were being published in the 1860s and 1870s, functioned as a popular science journal.[45]

Being published in Paris by an Ottoman intellectual, the various subjects treated in *Ulûm* reflect the imperial and cosmopolitan setting in which it was produced. Europe in the nineteenth century had many metropolises that reflected the impact of the growing colonial empires. Suavi and other Ottoman intellectuals and students were facing two challenges: first, criticizing their own states so that they reform, and thus, resist the imperial impingement of European colonial powers; and second, they had to respond to the criticisms of their cultures and histories by Orientalists. These two contemporary poles, together with the Islamic tradition, constituted the three sources that shaped *Ulûm*, and were being put in a dialectical relationship on its pages. Indeed, we can say that *Ulûm* in this regard is emblematic of the late Ottoman thought in general, as it also was influenced by contemporary European political and Orientalist discourses, Islamic-Ottoman discursive tradition, and contemporary Ottoman realities. These interwoven intellectual and social threads, which shaped Suavi's *Ulûm*, were a product of an increasingly interconnected world, owing to industrial developments including railways, telegraphy, steamship, and the printing press and periodical publishing platforms. Thus, Suavi's project of utilizing the press and the periodical platform appears to be twofold: the first was preserving existing knowledge [*hifẓ al-mawjūd*] and the second was seeking knowledge that was lacking [*ṣayd al-mafqūd*]. Analysis of select articles that take on issues of Orientalism, racism, writing, and the basis of Islamic legal theory, which appeared in *Ulûm*, evince the multiple threads that shaped Suavi's thought and the periodical. Furthermore, they show that *Ulûm* was as much a response to Orientalist discourses as it was shaped by them.

Ulûm as a product and critique of Orientalism and racism

Ali Suavi's first article in *Ulûm*, entitled "The Turk," is indicative of how the context of its production – Paris at the dawn of the Franco-Prussian War – stimulated its content (Figure 7.2). The article also indicates the major themes with which the journal would be preoccupied: science, religion, politics, and language. Given its importance, it deserves a close reading. In this article, Suavi is at pains to prove the Turkish contribution to political and scientific thought. It is clear that this article specifically, and many others in the following issues, are in reaction to European Orientalist and racist discourses, as evidenced by Suavi's first sentence: "There is an issue of race in Europe [*Avrupada râs meselesi var*], in other words, it is believed that in order to judge a people's [*kavim*, ethnic group or nation] capabilities and potentials, attention should be paid to the branch [of races] to which it belongs."[46] It is noteworthy that Suavi did not translate the word "race"; rather he provided the transliteration followed by a brief explanation of the word. This in itself is important, as it demonstrates that there were in fact unfamiliar themes and issues which Ottomans were encountering in European discourses, and thus these new relevant key concepts were transliterated into Ottoman Turkish.

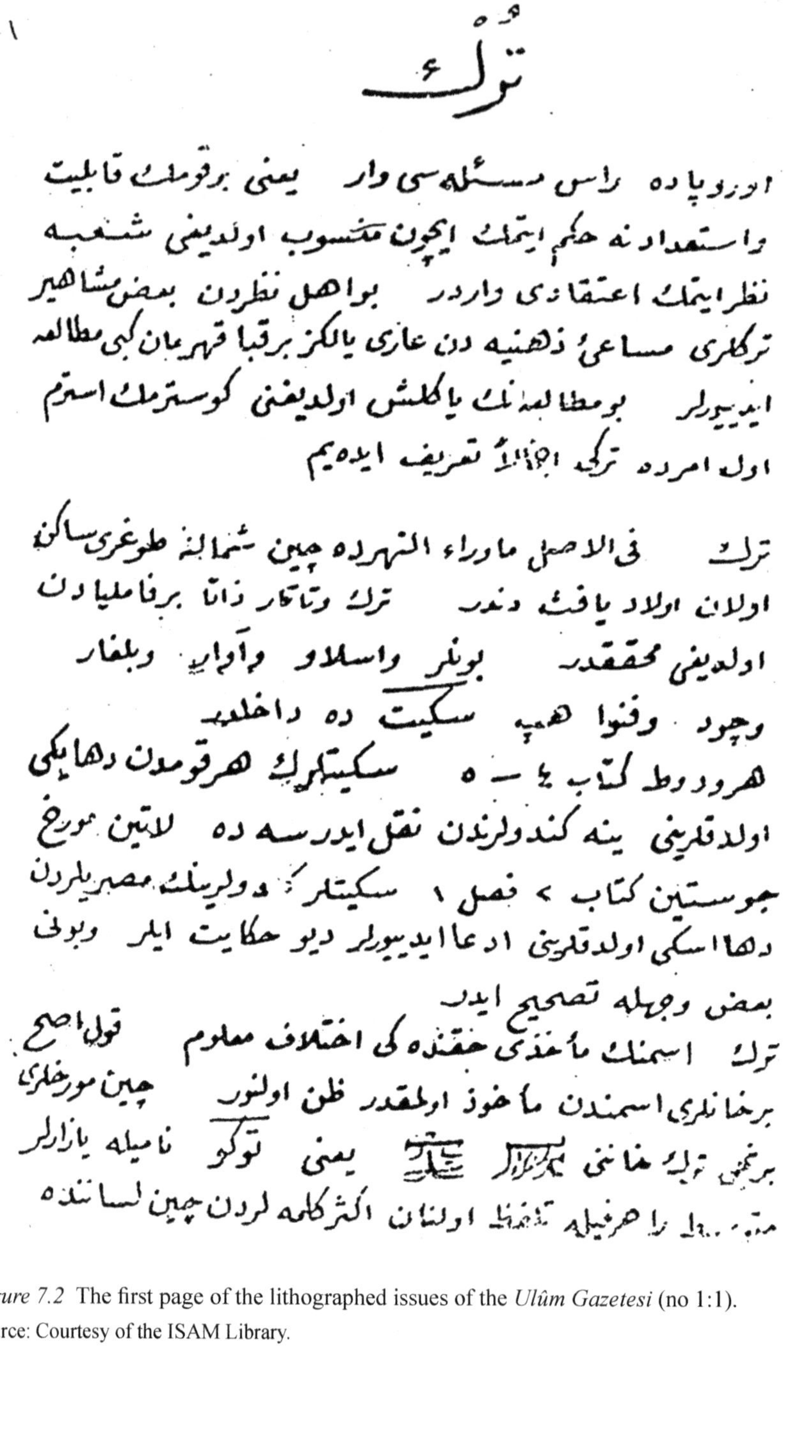

١

ترك

اوروپاده راس مسئله‌سی وار یعنی برقومك قابلیت
واستعدادنه حكم ایتمك ایچون منسوب اولدیغی شعبه
نظر ایتمك اعتقادی واردر بو اهل نظردن بعض مشاهیر
تركلری مساعیٔ ذهنیه دن عاری یالكز برقبا قهرمان كبی مطالعه
ایدییورلر بو مطالعه‌نك یاكلش اولدیغنی كوسترمك استرم
اول امرده تركه اجمالاً تعریف ایده‌یم

ترك فی الاصل ماوراء النهرده چین شمالنه طوغری ساكنه
اولان اولاد یافث دندر ترك وتاتار ذاتا برفاملیادن
اولدیغی محققدر بونلر واسلاو وآواریه وبلغار
وچود وفنوا هپ سكیت ده داخلدر
هرودوط كتاب ٤ — ٥ سكیتلرك هرقومدن دها ایكی
اولدقلرینی ینه كندولرندن نقل ایدرسه ده لاتین مورخ
جوستین كتاب ٢ فصل ١ سكیتلرك دولریلك مصریلردن
دها اسكی اولدقلرینی ادعا ایدییورلر دیو حكایت ایلر وبونی
بعض وجهله تصحیح ایدر
ترك اسمنك مأخذی حقنده كی اختلاف معلوم قول اصح
برخانلری اسمندن مأخوذ اولمقدر ظن اولنور چین مورخلری
برنجی ترك خاننی [illegible] یعنی توكو نامیله یازارلر
[illegible] ایله تلفظ اولنان اكثر كلمه لردن چین لساننده

Figure 7.2 The first page of the lithographed issues of the *Ulûm Gazetesi* (no 1:1).
Source: Courtesy of the ISAM Library.

Suavi's article, and his interest in the issue, stemmed from European intellectuals' representation of Turks as a rude heroic nation lacking in intellectual capabilities. An example of such a view, as reported by Suavi, was Alphonse de Lamartine's (d. 1869) depiction of Turks as a great Eastern nation that lacked a knowledge of politics.[47] Suavi set out to prove such representations wrong, beginning with the definition of the Turk. Suavi described the Turk as originating from the Transoxiana and being a descendent of one of Yafes' (Japheth) sons. Suavi added that Turks and Tatars were essentially of the same family. These groups, and for instance, Slavs, Avars, and Bulgars, were all contained under the label of Scythian. Suavi notes that the name "Turk" is of contested origin, though it is assumed to be the name of one of their khans. Suavi also mentions the Chinese reference to Toku, as the first Turkish khan, even reproducing the Chinese script and noting that it is a distortion of the name Turk. Then Suavi names various Turkish dynasties such as Ghaznavids, Seljukids, and Ottomans, and various Turkic peoples, including Ottomans, Uzbeks, and Turkmens.[48] From his references, it is clear that in this effort to define the Turk, Suavi relied on European sources, or as some scholars have asserted, perhaps solely relied on Arthur Lumley Davids's (d. 1832) *Grammar of the Turkish Language*.[49] It also appears that Suavi responded to Orientalist charges by adopting their very constructions of race as a category of thought. Thus despite his criticism of Orientalist charges against Turks, Suavi was himself participating in this discourse by reproducing Turks as a category within the frame of European, Orientalist racist thinking.

After defining the Turk, Suavi transitioned in the second section of this article to highlight Turkish intellectual contributions, beginning with a list of philosophers, theologians, and doctors of Turkish origin. This was intended as an account of ancient Turks that mostly overlapped with the pre-Ottoman Islamic period, such as al-Farabi (d. 950). Then, Suavi offered a report on the Ottomans' intellectual efforts. In this regard, he enumerated the names and works of famous Ottoman Turkish scholars in a number of fields, including practical philosophy, arithmetic, philosophy, literature, astronomy, natural philosophy, and law. Suavi interestingly added some information on the mastery of languages among Ottomans, and the existence of translations in Turkish from European languages such as French, Latin, and English. Similarly, Suavi noted that Europeans translated books from Arabic and Turkish into Latin, French, Italian, and English, citing some specific examples. Though this nuanced approach seems to challenge the dichotomy between East and West by registering a fluid transmission of knowledge between Turkish and European languages, this is not explicitly used to counter racist intellectual history. Although Suavi offers a more nuanced approach to Turkish contributions to science than what would have been available in Europe, he evidently also catered to the Orientalist-European structure of thinking, as evidenced by the apologetic nature of this account. In another series of articles appearing first in the third issue of the *Ulûm* and entitled "History of Thought," Suavi conveniently adopted the Orientalist division of East and West in writing a history of philosophy.[50]

At the same time as he criticized European Orientalists or intellectuals, Suavi also quoted them, sometimes even the same figures, as authorities to support his own arguments when their views reflected well on the Ottomans and Muslims. For instance, in responding to the widespread Orientalist and European depictions of the Ottoman Empire as politically dull and despotic, Suavi declared that the Ottoman government was established on good morals, and that there was nothing higher in the Ottoman government than the *shari'a* and the law [*kanun*].[51] Suavi asserted that there were no despots [*cebbâr*] in the Ottoman dynasty, defining a despot as a ruler who considered himself above the law. As further evidence to the antidespotic system of the Ottoman Empire, Suavi mentions the mechanism of *khal'* [deposing the ruler] in Ottoman law, which was exercised by the viziers and the community of scholars. He quotes François-René de Chateaubriand (d. 1848) to the effect that due to this system of dismissal the Ottoman State had a limited constitutional government.[52] Thus, Suavi's relationship with European scholars was not one of extreme rejection or imitation, rather it was one of critical engagement and appropriation.

From the preceding Turkish intellectual history, Suavi concluded that the Turks' power of cognition was capable of scientific effort. Unlike the Europeans who attributed developments in science during the Islamic period to Arabs, Suavi claimed that they were mostly a product of Turkish and Eastern people. Suavi alleged that Arabs for the most part became storytellers, and that there were few scholars among them, referring the reader to such classical sources as Ibn Khaldun's *Muqaddimah* and Katip Celebi's *Kasfh al-Ẓunūn*.[53] As another article by Suavi on the Turkish language indicates, he was responding to a claim by some Orientalists that Turks, Persians, and Indians did not contribute to science, since they lacked such contributions in their languages. There, Suavi differentiated between Arabic as a language of Arabs and Arabic as a language of Islam and the sciences.[54] The latter, he maintained, was constructed by non-Arabs as a medium for uniting various Muslim nations to avoid divisions and conflicts. Interestingly, Suavi asserted that contrary to common belief, it was not the Qur'an or religion that made Arabic the language of Islam (Muslims) and the language of sciences, but rather the idea of uniting various Muslim nations and ethnicities against the disruptive danger of national and ethnic divisions stimulated this development.[55]

All of these arguments indicate that Suavi was responding to Orientalist scientific racism,[56] which pitted various Eastern nations against each other, and caused rivalries and divisions among them. Once again, as he was challenging these generalizations, Suavi was also taking part in the very same racist and nationalist discourses.

Despite his efforts to defend Turkish and Ottoman history against European Orientalist representation, Suavi at the same time agreed that contemporary Ottomans were lagging behind in the sciences. He believed that was due to two mistakes. The first was the late introduction of the printing press. According to Suavi, the printing press was limited to publishing non-Islamic books, so as not to infringe on the work of calligraphers; thus, this technology faced a serious obstacle that prevented it from flourishing in the Ottoman Empire. This mistake,

Suavi maintained, gave Europeans an upper hand, surpassing the Ottomans by publishing and distributing more books through the employment of the printing press.[57] The second mistake, according to Suavi, was sending students to Europe, rather than opening schools in the Ottoman Empire and attracting foreign teachers to these schools.[58] In fact, the government did open schools in the Ottoman Empire, and Suavi himself was educated in one such school. Furthermore, those sent to Europe, in all likelihood, constituted a very small fraction of students educated in total, in comparison to those educated in the new schools within the Ottoman Empire. Suavi concluded the piece by prophesying that Oriental people would acquire European know-how and further improve it.

Suavi's relatively short but significant article on the Turk indicates that modern Ottoman periodicals on science emerged within an intellectual environment that had to also contend with European racist, nationalist, and Orientalist discourses. In criticizing Orientalist representations of Islamic and Arabic intellectual history, some Ottoman Turkish scholars, including Suavi, were also adopting those nationalist discourses and hence producing an intellectual history that was mirroring the West and would influence the later historiography as well. Furthermore, this particular article, which is essentially an introduction of the brand new periodical, reflects Suavi's struggles both at home and abroad: resisting European colonial and Orientalist discourses while calling for reform in the Ottoman Empire. In that regard, the inaugural issue and this very first article, "The Turk," indicate that this periodical will be a vehicle for propagating the sciences, calling for reform in various areas including language and politics, and educating Ottomans with regards to certain technological developments, such as printing.

Ulûm as an instrument of moderate and radical calls for reform

In addition to the critiques of Orientalism, Suavi also discussed in *Ulûm* contemporary developments in the Ottoman Empire. Among these commentaries, Suavi's articles on the alphabet and writing and religious reforms further reveal how the fact of his physical location in Europe and his disagreements with other Young Ottomans informed the perspective presented in the journal on reforms in the Ottoman Empire. While finding some reform projects to be flighty, Suavi considered it apt to radically break with some of the most fundamental elements of Islamic intellectual tradition, such as the traditional legal theory.

A piece concerning the alphabet and writing that appeared in the fourth issue used a conversational format (employing question-and-answer style) in order to discuss the benefits of reforming the alphabet. An unnamed questioner – presumably another Young Ottoman fictionalized by Suavi himself – commends a brief history of the alphabet and writing provided by Suavi in an earlier article. The questioner concludes that from Suavi's argument readers would no doubt be persuaded that it would not be a bad innovation [*bid'at-ı seyyie*] if the Ottomans reformed [*ıslah*] their script [*hat*], and that in fact it would be a good innovation [*bid'at-i hasene*] based on the positive example set by certain predecessors [*eslaf*].

In response to the questioner, Suavi states that "no doubt reforming the script is a good deed [*emr-i hasene*]. [But that] it is better not to use the phrase *bid'at* [innovation] in regards to such matters since *bid'at*, in the terminology of shari'a, refers to forging something that is not within religion [*din*]."[59] At the same time, Suavi maintains that such an issue as writing has nothing to do with religious matters [*umûr-ı diniye*], but is rather within the secular realm [*umûr-ı dünyeviye*]. Suavi justifies the discussion by quoting the Prophet as stating, "You are more knowledgeable in your worldly matters."[60] This is an allusion to an instance in which the Prophet tells the farmers that because they know more about agriculture than he does, his discourse on such matters came not in the form of revelation, but merely his opinion as a fellow human being.

Upon receiving this response, the questioner invites Suavi to destroy the alphabet entirely, to which Suavi inquires after his intent. The interlocutor wishes to replace the Arabic script in order to make it easier for people to grasp it more quickly and with ease, just as, according to him, Europeans were able to learn the Roman alphabet in ten days. Thus the interlocutor claims that a reform to the script of the alphabet would make it easier to "spread sciences" [*neşr-i ulûm ve maârif*].[61] Suavi confirms that there were indeed other people who were of the same thought but cautions that a desire to destroy something is either due to imitation [*taklid*] or whim [*heves*] and rejects the claim that it would be easier to learn the European alphabet. Citing English as an example, Suavi notes that the same vowel "a" is pronounced differently in the words, "far, fat, fate, fall, warm, arise," which makes the language challenging for students to master.[62] Suavi thus portrays himself as a moderate synthesizer of the old and the new.

While the piece on reforming script was a call for moderation, other pieces entertain a radical break with some of the traditional Ottoman-Islamic practices. In this regard, among *Ulûm*'s most striking pieces are the articles related to religious reform, though they do not occupy much of the periodical. In an article entitled, "Semi jurist (theologian) destroys religion [*yarım fakih din yıkar*]," Suavi looks at the potentially harmful influence of theologians in the context of all religions, and exhibits a preference for rational and universal principles that was quite modern.[63]

In the succeeding article, unapologetically entitled "Lets reject laws that are based on literary studies," Suavi bolstered his criticism of *uṣūl al-fiqh* [the principles of law], because they were based on the Arabic linguistic sciences. According to Suavi, the interpretation of religious texts based on such contested issues of language did not provide a firm ground for a durable and powerful law, and thus politics founded on such law was inherently weak. Suavi would have preferred that an earlier generation of Muslim scholars had not mixed worldly and otherworldly issues by including both of them within law [*fiqh*] books. This separation, Suavi contends, would have made it possible to establish politics based on principles derived from such disciplines as geography, political economy, and ethics, making good governance of the state possible and more rational.[64] Notably, these are disciplines that held strong in Europe, especially with regards to political economy; thus, his criticism of Islamic law and legal theory was informed by contemporary European approaches to politics.[65] Indeed, Suavi further claims that while Europeans were studying astronomical principles and the causes of wealth and civilization, Easterners were busy with such matters as *abjad* calculations for the purpose of interpreting the Qur'an.[66] From his geographical position in France, it is clear that Suavi was influenced by European examples, and his ideas for the transformation of Islamic law were motivated by a desire to forge a new kind of European-modeled jurisprudence and political theory.

Suavi, aware of the power of previous generations in reconstructing religion, proposes radical reforms in Islamic law and legal theory by calling for the separation of worldly and otherworldly matters, of religious law and the principles of politics. Instead of relying on the older Islamic legal theory, which was based on Arabic linguistics, he proposes – well in line with the developments in Europe at that time – that law be based on universal and rational principles, and that politics should be based on political economy, geography, and ethics. Hence, we find that the content of this periodical reflects concerns of the late Ottoman and European setting, while the author takes an uneven stance regarding various issues.

Conclusion

As access to printing technologies expanded in the nineteenth century, and the world became increasingly interconnected through faster modes of travel and communication, intellectuals and journalists who were barred from expressing their views could seek alternative dominions whence they would publish freely. Encountering such a situation in the Ottoman Empire, Ali Suavi, an intellectual, journalist, and political activist, escaped to Europe and continued his publishing activities there. His journal of sciences, *Ulûm Gazetesi*, which was published from Paris, reflects the possibilities as well as vulnerabilities surrounding early Ottoman periodicals, especially those published abroad. As Suavi himself acknowledged and the history of *Ulûm* attests, a printing enterprise required more than a secure space, but also financial stability, a communicable language, and readers. Of all these three elements, Suavi could obtain the latter two, by writing in an accessible manner that could appeal to a large audience. He also had wide-ranging and provocative ideas that also could ensure consistent readership.

Despite being published by a rather skilled intellectual who knew how to navigate this intricate enterprise, *Ulûm Gazetesi* nevertheless faced two main challenges: lack of stable financial resources and an editorial staff. *Ulûm Gazetesi* was the work of one man, Ali Suavi, the sole publisher-author. This brought with it a struggle to maintain a level of robustness in the articles. Though Suavi managed to find things to put into ink and press onto paper, reading *Ulûm Gazetesi* one gets a sense that much was included as filler, rather than intended for careful analysis. This was also reflected in ad hoc changes to the subjects included in the journal, as many notices have indicated. However, the journal did include something for multiple audiences in Europe and in the Ottoman Empire.

The content of the *Ulûm* was shaped by three major influences: contemporary European discourses, contemporary Ottoman developments, and Islamic intellectual tradition. Since the spread of the printing press and periodicals in the Ottoman

Empire, it was very difficult for a modern Muslim intellectual to completely ignore Orientalist ideas in an increasingly global world. Even though Suavi wrote the content for the journal in Turkish, and his main audience consisted of Ottoman readers in in the homeland as well as émigrés in Europe, the journal's place of publication – Paris in 1869–1870 – enabled him to have more frequent interactions with Orientalists and European intellectuals. Therefore, many articles in the *Ulûm* contain responses to or influences from the European setting. However, despite his close engagement with European intellectuals, Suavi, and the journal by no means were mere transmitters of European ideas. Rather the interaction was one of critical evaluation and appropriation. Suavi's engagement with the Islamic intellectual tradition also parallels his reactions to contemporary European discourses, as he at times defended this tradition against Orientalists, while simultaneously not hesitating to criticize the contemporary condition of the Ottoman state. Suavi was more critical of the Ottoman government than anything else. Overall, in the pages of *Ulûm*, we find a hybrid discourse that was aimed at reform in the Ottoman Empire, and defense against European Orientalist misrepresentations, while drawing on a history of intellectual thought from these resources.

Rather than being a peculiar case, *Ulûm Gazetesi* embodies the emergence of a genre of periodicals in exile around the world. Therefore, this periodical lends itself for a comparative study of the uses of print in exile, and the unique perspectives they can provide on politics at home and developments abroad. In that regard studying this Ottoman Turkish popular journal of science sheds as much light on matters Ottoman as it does on matters European. Suavi's journal of *Ulûm* shows how some non-European publisher-authors situated themselves as reformists and resisters at once, thus challenging the popular Eurocentric narratives of the modern period.

Notes

1 I am grateful to Meghan Forbes for organizing a conference on the printing press and periodicals ("Summoning the Archive: A Symposium on the Periodical, Printed Matter, and Digital Archiving"), where I presented a paper from which this chapter has grown. I would like to also thank Meghan Forbes and Ercüment Asil for providing me with helpful comments and constructive criticisms on an earlier draft of this chapter.

2 For the impact of print culture on the West, see for instance the following influential studies: Elizabeth L. Eisenstein, *The Printing Press as an Agent of Change: Communications and Cultural Transformations in Early-Modern Europe*, vols. 1 and 2 (New York: Cambridge University Press, 2009); Benedict R. O'G. Anderson, *Imagined Communities: Reflections on the Origin and Spread of Nationalism* (London: Verso, 1983).

3 The Ottoman case is peculiar in that it adopted the widespread use of the printing press rather late. This is much discussed in the secondary literature. For a discussion and relevant references see Ami Ayalon, *The Arabic Print Revolution: Cultural Production and Mass Readership* (Cambridge: Cambridge University Press, 2016).

4 For a collection of Ottoman regulations concerning the press and publications see Server Rifat İskit, *Türkiyede matbuat rejimleri* (İstanbul: Ülkü, 1939).

5 For a recent study of popular science periodicals that included alternative paths of modernization, see Ercüment Asil, "The Pursuit of the Modern Mind: Popularization of Science, the Development of the Middle Classes, and Religious Transformation in the Ottoman Empire, 1860–1880" (Ph.D. Diss. University of Chicago, Chicago, 2017).

For a particularly contentious case of printing and its impact in Turkey, see M. Brett Wilson, *Translating Qur'an in an Age of Nationalism: Print Culture and Modern Islam in Turkey* (Oxford: Oxford University Press, 2014).

6 Young Ottomans were an oppositional group that emerged in the 1860s. The most prominent members included Namık Kemal, Mustafa Fazıl Pasha, Ziya Pasha, and Ali Suavi. For a study of these prominent members and the Young Ottoman movement in general, see Şerif Mardin, *The Genesis of Young Ottoman Thought* (Princeton, NJ: Princeton University Press, 1962).

7 Ali Suavi, "Bab-ı Âlî İdaresi Kaidelerinden Birkaçı," *Ulûm* 2, no. 14 (1870): 830.

8 For such a portrayal of Suavi, see the section "Ali Suavi: the Zealot," in Mardin, *The Genesis of Young Ottoman Thought*. For a more positive representation, refer to Kemal Karpat, *Politicization of Islam* (New York: Oxford University Press, 2001). Representation of Suavi in the literature is quite similar to that of another leading Islamist and exploiter of this new medium of the printing press and periodicals, namely Jamāl al-Dīn al-Afghāni (d. 1897).

9 Suavi, *Ulûm* 1, no. 1 (1869).

10 In the announcement, Suavi was advertising his works, beginning with the first volume of *Ulûm*. Suavi listed its price as 20 francs. See *Ulûm* 2, no. 11 (December 20, 1869): 679–680.

11 The advertisements that describe the journal come before the content page. Since I have only been able to access the photocopied and collated versions of the first volume (in Atatürk Kitaplığı of İstanbul) rather than the separate issues, I am not sure if they belong to the issue preceding the ad or the one succeeding it. I have also made use of the copies available at İSAM library, which also seem to be based on the above mentioned copy, yet they do not include the ads preceding the content pages.

12 In a note in the sixth issue, Suavi claims that letters and other materials necessary for the press are ready; however, he will not letterpress print so that the costs the customers have to bear remain the same. In other words, if we are to believe Suavi, he was postponing letterpress printing as it would cost more for the customers, and thus, logically, for himself as well. Suavi, *Ulûm* 1, no. 6 (1869).

13 Although lithography seems to have been the preferred method of printing among the Muslims of Persia and the Subcontinent, as pointed out by Ian Proudfoot, it remained in the shadow of the letterpress printing in the core Ottoman lands, as also suggested by Suavi's reluctant use of this method. For a comparison between two methods of printing in the Arabic script, see Ian Proudfoot, "Mass Producing Houri's Moles," in *Islam: Essays in Scripture, Thought and Society: A Festschrift in Honour of Anthony H. Johnes*, Anthony H. Johns and Peter G. Riddell (Leiden: Brill, 1997), 161–184.

14 For a study of this encyclopedia together with others in the context of late Ottoman modernization, see Jutta R. M. Çıkar, *Fortschritt durch Wissen: Osmanisch-türkische Enzyklopädien der Jahre 1870–1936* (Wiesbaden: Harrasowitz Verlag, 2004).

15 Ali Suavi, *Ulûm*, no. 24 (August 16, 1870).

16 Ali Suavi, "İlan," *Muvakkaten Ulûm Gazetesi Müşterilerine*, no. 1 (September 30, 1870): 1.

17 Ibid., 1–3.

18 Suavi, "İlan," *Ulûm* no. 21 (1870): 1276. In his autobiography, Suavi mentions that he would send one hundred copies of *Le Mukhbir* to Istanbul, for circulation in the city. Considering this number for a political newspaper, subscription of *Ulûm* could be lower. *Ulûm* 2, no. 15 (1870): 927–928.

19 The financial struggles can be seen from Suavi's desperate attempts in the twenty-third issue at increasing the number of subscribers. Suavi promises to those who subscribe for one year, a free copy of a prospective weekly journal entitled *Mehdi*. Suavi advertises this as a not-for-sale journal that would focus on news related to the state administration, which would be given to *Ulûm* subscribers. Separately, Suavi promises to give another prospective journal of his, which he calls *Bab-ı Âlî* (Sublime Porte),

to subscribers. These two journals are described similarly, with the exception that the former would be a monthly. Suavi also promises to give free copies of publications of an Islamic Society that he claims to have established. I find this advertisement a desperate call for new subscriptions as Suavi promises to give intangible objects at the time of the announcement. See Suavi, "İlan," *Ulûm* no. 23 (1870): 1337.

20 See, for instance, backmatter of *Ulûm*, no 24, which also includes a table of contents in French.

21 Ali Suavi, "Kamusu'l-Ulûm ve'l-Maarif Zeyli," *Ulûm* no. 24 (1870): 1383.

22 Ali Suavi, "Yeni Osmanlılar Tarihi," *Ulûm* no. 15 (1870): 892.

23 Hüseyin Çelik, *Ali Suavi ve Dönemi* (Ankara: T.C. Kültür Bakanlığı, 1993), 24–25.

24 For Suavi's autobiography see Suavi, "Yeni Osmanlılar Tarihi," 892–932. For more on Suavi's biography see Hüseyin Çelik, *Ali Suavî Ve Dönemi* [Ali Suavi and His Times] (Cağaloğlu, Istanbul: İletişim, 1994), 41–112. For further literature on Suavi's biography and an assessment of his work see the recent dissertation by Aaron S. Johnson, "A Revolutionary Young Ottoman: Ali Suavi (1839–1878)" (Ph.D. Diss., McGill University, Montreal, 2012).

25 Suavi notes that he presented a copy of his first composition, a commentary on a creed, to one of the teachers in the Bayezid Mosque in Istanbul, indicating that he probably learned Arabic and studied the classical *medrese* curriculum with such teachers. Suavi, "Yeni Osmanlılar Tarihi," *Ulûm* 2, no 15 (1870): 894. In the first article of *Ulûm*, which will be analyzed shortly, Suavi mentions the unmatched educational role mosques played in the Ottoman Empire. I take this as another cue that they played such a role in his own educational formation.

26 Suavi, "Yeni Osmanlılar Tarihi," *Ulûm* 2, no. 15 (1870): 892–893.

27 Ibid., 895. In a notice Suavi mentions that he taught geometry and land measurements at Bağhane Medresesi in Filibe. See "İlan," *Ulûm* 1, no. 7 (1869): 419.

28 Çelik, *Ali Suavi ve Dönemi*, 5–7.

29 In addition to mentioning his father's unwavering stance for justice, Suavi noted the knowledge of *ḥadīth* as an influence on his political formation. He had memorized books on this literature which basically narrate the Prophet's discourse and way of life. The hadīth literature that Suavi studied includes Suyūtī's (d. 1505) small collection (*Jāmi' al-Saghīr*), and Minawi's commentary (*Fayḍ al-Qadīr Sharḥ al-Jāmi' al-Saghīr*) on that collection. Suavi notes that he memorized Daylamī's collection (*Firdaws*) as well, and summarized Bukhārī's *Sahih*. Suavi, "Yeni Osmanlılar Tarihi," *Ulûm* 2, no. 15 (1870): 896.

30 Suavi, "Yeni Osmanlılar Tarihi," *Ulûm* 2, no. 15 (1870): 896–897.

31 In fact, the relationship between preaching and printing was made clear in various writings on the impact of periodicals, such as *Mecmua-i Fünûn* [Journal of Sciences] and *Tercümân-ı Hakikât* [Translator of the Truth].

32 Suavi, "Yeni Osmanlılar Tarihi," *Ulûm* 2, no. 15 (1870): 908.

33 Ibid., 910. Suavi sought to emphasize his detachment from Fazil Pasha by stating that he had at the same time turned down a request to meet from his servant [*kethüda*].

34 Suavi, "Yeni Osmanlılar Tarihi," *Ulûm* 2, no. 15 (1870): 910–911.

35 Ibid., 916. Another lure was joining Mustafa Fazil's reformist calls as expressed in a letter that was first published in European newspapers, and then translated to Turkish and lithographed. Suavi read this letter while in exile. Suavi, "Yeni Osmanlılar Tarihi," *Ulûm* 2, no. 15: 917. For a copy of the letter, see Mustafa Fazil Pasha, *Paristen Bir Mektup* [A Letter From Paris] (Dersaadet: Artin Asadoryan Matbaası, 1326).

36 Author Anon, "(Muhbir) doğru söylemek yasak olmayan bir memleket bulur yine çıkar," *Le Mukhbir* no. 1 (August 31, 1867): 1.

37 The word *ulema* traditionally was used to refer to learned scholars that were educated in the religious and rational sciences. In the Ottoman Empire it referred to scholars who were appointed in the judiciary or schools. In the nineteenth century it was increasingly being used to refer to scientists.

38 Suavi, "İlan," *Ulûm* 1, no. 3 (1869): 162.
39 Suavi, "İlm-i Hesap," *Ulûm* 1, no. 4 (1869): 183–213.
40 Suavi, "İlan," *Ulûm* 1, no. 8 (1869): 478.
41 Interestingly, Mardin, *The Genesis of Young Ottoman Thought*, dubbed Suavi's style as being primitive, without considering his conscious choice here. Mardin's negative portrayal of Suavi might be due to the influence of Suavi's intellectual nemesis, Namık Kemal's pronouncements.
42 Suavi, "Tembih," *Ulûm* 1, no. 8 (1869): 466.
43 Suavi, "Tab‘," *Ulûm* 2, no. 15 (1870): 879.
44 Suavi, "İlan," *Ulûm*, no. 21: 1275.
45 For similar journals in the Ottoman Empire that are considered as popular science journals, see Asil, "The Pursuit of the Modern Mind," 48–52.
46 Suavi, "Turk," *Ulûm* 1, no. 1 (1869): 1.
47 Suavi, *Ulûm* 1, no. 1 (1869): 15. I have not been able to pin down the exact quote from Lamartine. However, it should be said that he wrote a voluminous history of Turkey. See Alphonse de Lamartine, *Histoire de la Turquie* (Paris: Lecou-Pagnerre, 1855).
48 Ibid., 1–2.
49 See Ercüment Kuran, "Ottoman Historiography of the Tanzimat Period," in *Historians of the Middle East*, ed. B. Lewis and P. M. Holt (London: Oxford University Press, 1962), 428. For the selective approach of Suavi to Lumley's *Grammar* cf. Tanpınar, *19. Asir Turk Edebiyatı*, 220; Arthur L. Davids, *A Grammar of the Turkish Language: With a Preliminary Discourse on the Language and Literature of the Turkish Nations, a Copious Vocabulary, Dialogues, a Collection of Extracts in Prose and Verse, and Lithographed Specimens of Various Ancient and Modern Manuscripts* (London: Parbury & Allen, 1832). For a reference to A. L. Davids, see Suavi, *Ulûm* 1, no. 1 (1869): 7–8, 11.
50 Suavi, "Tarih-i Efkâr," *Ulûm* 1, no. 3 (1869): 105.
51 Suavi, *Ulûm* no. 1 (1869): 15. For a study on Enlightenment representations of Turks, see Aslı Çırakman, "From Tyranny to Despotism: The Enlightenment's Unenlightened Image of the Turks," *International Journal of Middle East Studies* 33, no. 1 (2001): 49–68.
52 Suavi, *Ulûm* 1, no. 1 (1869): 15.
53 See Ibn Khaldun, *Al-Muqaddima*, v. 3: 229–232; Katip Çelebi, *Kashf al-Zunūn*, v. 1: 54; cf. ibid., 55, where it is claimed that Arabs were not taught any philosophical sciences.
54 Suavi, "Lisan ve Hatt-ı Türki," *Ulûm* 1, no. 3 (1869): 124–125.
55 Ibid., 129.
56 For a now famous discussion and analysis of European scientific racism, see Edward Said, *Orientalism* (New York: Pantheon Books, 1978). Another piece in *Ulûm*, entitled "European Objection" defends Islam's ability to reform against European bias, and at the same time it criticizes Western educated Ottomans' inability to instigate home-grown reform. See Suavi, "Avrupa İtirazı," *Ulûm* 2, no. 19 (1870): 1185–1191.
57 Suavi, *Ulûm* 1, no. 1 (1869): 17. For a separate article discussing the issue of the printing press in the context of Ottoman publications in the twelfth century AH (roughly eighteenth century CE), see *Ulûm* 1, no. 8 (1869): 458–465.
58 Suavi, *Ulûm* 1, no. 1 (1869): 17.
59 Suavi, "Lisan ve Hatt-ı Türki," *Ulûm* 1, no 4 (1869): 219.
60 Ibid.
61 Ibid., 220.
62 Ibid., 221. The example from English was used in the imperial dictionary. Perhaps it was one of the examples through which Suavi learned the nuances of English pronunciation. John Oglivie, *The Imperial Dictionary of the English Language*, vol. 4 (London: Blackie, 1883), xviii.
63 Suavi, "Yarım Fakih Din Yıkar," *Ulûm* 2, no. 17 (1870): 1025–1027.
64 Suavi, "Edebiye Üzre Bina Kılınan Ahkamı Red Edelim," *Ulûm* 2, no. 17 (1870): 1028–1030.

65 Suavi's discourse and the late Ottoman governmental practices suggests that the domain of politics in the late Ottoman period was expanding due to the emergence of political economy.

66 Suavi, "Edebiye Üzre Bina Kılınan Ahkamı Red Edelim," *Ulûm* 2, no. 17 (1870): 1036.

Bibliography

Ahmed Cevdet Pasha. *Terceme-i Mukaddeme-i İbn Haldun*. Istanbul: Takvimhane-i Amire, 1858.

Anderson, Benedict R. O'G. *Imagined Communities: Reflections on the Origin and Spread of Nationalism*. London: Verso, 1983.

Asil, Ercüment. "The Pursuit of the Modern Mind: Popularization of Science, the Development of the Middle Classes, and Religious Transformation in the Ottoman Empire, 1860–1880." PhD Diss., Chicago University, Chicago, 2017.

Ayalon, Ami. *The Arabic Print Revolution: Cultural Production and Mass Readership*. Cambridge: Cambridge University Press, 2016.

Berkes, Niyazi. *The Development of Secularism in Turkey*. Montreal: McGill University Press, 1964.

Çelik, Hüseyin. *Ali Suavî ve Dönemi*. Cağaloğlu, Istanbul: İletişim, 1994.

Çırakman, Aslı. "From Tyranny to Despotism: The Enlightenment's Unenlightened Image of the Turks." *International Journal of Middle East Studies* 33, no. 1 (2001): 49–68.

Davids, Arthur L. *A Grammar of the Turkish Language: With a Preliminary Discourse on the Language and Literature of the Turkish Nations, a Copious Vocabulary, Dialogues, a Collection of Extracts in Prose and Verse, and Lithographed Specimens of Various Ancient and Modern Manuscripts*. London: Parbury & Allen, 1832.

Eisenstein, Elizabeth L. *The Printing Press as an Agent of Change: Communications and Cultural Transformations in Early-Modern Europe*, Vol. 1 and 2. New York: Cambridge University Press, 2009.

İskit, Server R. *Türkiyede Matbuat Rejimleri*. Istanbul: Ülkü matbassi, 1939.

Johnson, Aaron S. "A Revolutionary Young Ottoman: Ali Suavi (1839–1878)." PhD Diss., McGill University, 2012.

Karpat, Kemal H. *The Politicization of Islam: Reconstructing Identity, State, Faith, and Community in the Late Ottoman State*. New York: Oxford University Press, 2001.

Kâtip Çelebi. *Kitab Kashf al-Ẓunūn*. Lebanon: Dar al-Kotob al-Ilmiya, 2008.

Khaldun, Ibn. *Al-Muqaddima*. Edited by Abdesselam Cheddadi. Al-Dār al-Baydā', 2005.

Kuran, Ercüment. "Ottoman Historiography of the Tanzimat Period." In *Historians of the Middle East*, edited by B. Lewis and P. M. Holt. London: Oxford University Press, 1962.

Mardin, Şerif. *The Genesis of Young Ottoman Thought: A Study in the Modernization of Turkish Political Ideas*. Princeton Oriental Studies v. 21. Princeton, NJ: Princeton University Press, 1962.

Mustafa Fazil Pasha. *Paristen Bir Mektup*. Dersaadet: Artin Asadoryan Matbaası, 1326.

Oglivie, John. *The Imperial Dictionary of the English Language*, Vol. 4. London: Blackie, 1883.

Palabiyik, Nil. "An Early Case of the Printer's Self-Censorship in Constantinople." *The Library: the Transactions of the Bibliographical Society* 16, no. 4 (2015): 381–404.

Proudfoot, Ian. "Mass Producing Houri's Moles." In *Islam: Essays in Scripture, Thought and Society: A Festschrift in Honour of Anthony H. Johnes*, edited by Anthony H. Johns and Peter G. Riddell. Leiden: Brill, 1997.

Said, Edward W. *Orientalism*. New York: Pantheon Books, 1978.

Suavi, Ali, ed. *Le Mukhbir*, no. 1 (August 31, 1867).

Suavi, Ali. "Avrupa İtirazı." *Ulûm Gazetesi* 2, no. 19: 1185–1191.

Suavi, Ali. "Bab-ı Âlî İdaresi Kaidelerinden Birkaçı." *Ulûm Gazetesi* 2, no. 14: 827–830.

Suavi, Ali. "El-Hakimu Huvellâh." *Ulûm Gazetesi* no. 3: 135–161.

Suavi, Ali. "Edebiye Üzre Bina Kılınan Ahkamı Red Edelim." *Ulûm Gazetesi* 2, no. 17: 1018–1048.

Suavi, Ali. "Hülasatü'l-Haber fi'l-İlm ve'l-Eser der Karn-ı Sâni Aşer." *Ulûm Gazetesi* 2, no. 8: 426–465.

Suavi, Ali. "Kudret-i Siyasiye der Düvel-i İslamiye." *Ulûm Gazetesi* 2, no. 16: 981–1000.

Suavi, Ali. "Lisan ve Hatt-ı Türki." *Ulûm Gazetesi* 1, no. 2–3: 69–78, 115–134.

Suavi, Ali. "Tarih-i Efkâr." *Ulûm Gazetesi* 1, no. 3: 105–115.

Suavi, Ali. "Tab." *Ulûm Gazetesi* 2, no. 15: 879.

Suavi, Ali. "Türk." *Ulûm Gazetesi* 1, no. 1: 1–17.

Suavi, Ali. "Yarım Fakih Din Yıkar." *Ulûm Gazetesi* no. 17: 1025–1027.

Suavi, Ali. "Yeni Osmanlılar Tarihi." *Ulûm Gazetesi* no. 15: 892–932.

Tanpınar, Ahmet Hamdi. *XIX. Asır Türk Edebiyatı Tarihi*. Istanbul: İbrahim Horoz Basımevi, 1956.

Tekin, Kenan. "Reforming Categories of Science and Religion in the Late Ottoman Empire." PhD Diss., Columbia University in the City of New York, 2016.

Wilson, M. Brett. *Translating the Qur'an in an Age of Nationalism: Print Culture and Modern Islam in Turkey*. London: Oxford University Press, in association with the Institute of Ismaili Studies, 2014.

Yalçınkaya, M. Alper. *Learned Patriots: Debating Science, State, and Society in the Nineteenth Century Ottoman Empire*. Chicago: The University of Chicago Press, 2015.

8 Everyday printed matter

Kurt Schwitters's experimental typography[1]

Hannah Pröbsting

In the late 1920s, the aesthetics of the printed page were undergoing radical and rapid changes in Germany. The early years of the twentieth century had brought with them an increased interest in both the typography and layout of all types of publications, from text-based books to invitation cards and advertising posters. Yet this change came not from within the profession, but rather from the outside, with many avant-garde artists creating innovative new typefaces and experimenting with design, based on a new understanding of modern perception. The result was an aesthetic that we now recognize as "New Typography," which sparked a fundamental rethinking of how everyday printed matter could be used as a means of visual communication. One of the leading figures in this change was Kurt Schwitters (1887–1948) – a German avant-garde artist, writer, and graphic designer. Schwitters, however, was not content to just design experimental new typefaces – although he also did that – instead, he focused on the creation of an entirely new way of writing. What Schwitters envisaged was a rationalized script that would be compatible with twentieth-century life, and in particular, the speed of the modern city. By first of all examining Schwitters's *Systemschrift* [systematic script] and contextualizing it within broader typographical changes at the time, this chapter examines a range of everyday printed matter designed by Schwitters that used the *Systemschrift* and other related scripts. In doing so, it addresses questions of temporality and rationalization in relation to typography and highlights how Schwitters leveraged his commercial commissions to fulfill the avant-garde aim of bringing art into everyday life.

New Typography and the poster

Since the end of the nineteenth century, theorists and typographers had been battling over which kind of script – the ornate *Fraktur* (Gothic) or the simpler *Antiqua* (Latin) – was most suitable for texts published in German.[2] While the *Antiqua-Fraktur* debate continued to be a heated discussion in Germany throughout the 1920s, the years of the Weimar Republic (1919–1933) saw a marked shift towards *Antiqua.* This change was part of a broader aesthetic move from ornamentation to functionalism that can be seen across many disciplines at the time, with artists and designers alike rejecting the intricate designs of the Arts and Crafts and

Art Nouveau movements in favor of a more minimalist approach. In avant-garde circles this was propagated by the Bauhaus, whose professors and students – most notably Herbert Bayer, László Moholy-Nagy, and Josef Albers – made major contributions to typographic experimentation.[3]

It is in this context that the concept of New Typography was born. The New Typographers comprised a number of individuals, mostly working independently, who together had perhaps the greatest influence on the widespread shift from *Fraktur* to *Antiqua* and the popularization of sans-serif scripts in Germany.[4] Although a formal group was never founded, many of these artists equated their work with the term New Typography, thereby creating an unofficial movement.[5] Julia Meer defines them as a group of around twenty avant-garde artists, who, in addition to designing new typefaces, also contributed to a body of theoretical texts on the subject. Dynamism, stark contrasts, asymmetry, and the use of bold lines and geometric forms were the most important formal characteristics of this new aesthetic, she notes, stemming from notions of economy, functionality, modern perception, the psychology of advertising, standardization, and technology.[6]

The aesthetic shifts that were taking place in typography also had an impact beyond the letterforms, seeping into the design of the printed page itself. Instead of the elaborately decorated posters of the early twentieth century, the emphases of the New Typographers prompted them to create bold, striking images that contained the minimum of information to facilitate reading at high speed.[7] Perhaps one of the reasons that the New Typographers had such a significant influence on poster design was that its practitioners were not trained typographers, but rather prominent avant-garde artists from across Europe, including Albers, Bayer, Moholy-Nagy, Schwitters, Theo van Doesburg, El Lissitzky, and Alexander Rodchenko. Indeed, with the sole exception of Jan Tschichold, each of the New Typographers was self-taught, coming to the fields of typography and graphic design from the fine arts.

The transformations that were taking place in the typographical world were therefore deeply rooted in those happening in the fine arts, and strengthened by the networks that existed within them. For while each of the New Typographers was working independently, they were in no way isolated, since they were already connected through the strong networks of the European avant-gardes. Indeed, many of the New Typographers were working together on creative, typographic projects, such as van Doesburg and Schwitters's children's book, *Die Scheuche: Märchen*, written together with Käte Steinitz, which was illustrated with images composed entirely of letters and other typographical forms (lines, punctuation, etc.).[8] Similarly, Schwitters's commercial collaboration with El Lissitzky on advertisements for Pelikan, an international ink firm based in Hannover, led to striking posters that highlighted the innovative ways typography could be used on the page. In a different vein, Tschichold included many different examples of the work of New Typographers in his seminal book *The New Typography: A Handbook for Modern Designers*, thus creating a direct line between his own practice and that of his colleagues.[9] In each of these instances – and there are many others besides – the posters and books themselves become the publishing platform by which collaboration played out, and through which networks were established and strengthened.

In particular, the poster became a site of multiple levels of communication, taking on a similar role for the New Typographers that the little magazines and avant-garde journals held for networks of the avant-gardes. On the most fundamental level, New Typographers' posters conveyed information about the product or event they are advertising, yet on another, they are a space in which – as with the journals – conversations about typography were carried out and displayed. An important difference between these two examples, however, is that the everyday nature of these posters meant that they had a far broader reach. Displayed in public spaces, free for all passersby to see, rather than restricted to the pages of subscription-only journals, posters offered New Typographers the opportunity to magnify the impact that many of them had previously had in these more elite publishing venues.

An illustrative example of this utilization of posters can be found in the work of the *Ring: "neue Werbegestalter"* [Ring of New Advertisers] – a loose association of typographers and designers, founded by Schwitters in 1927. It sought to "to promote and popularize modern graphic design" by touring exhibitions on typography across Germany and in neighboring countries.[10] While some of the individual works on view varied from venue to venue, the concept of the *neue Typografie* [New Typography] exhibition remained the same: to showcase the typographical work of the Ring of New Advertisers' members – all of whom were prominent New Typographers.[11] This exhibition displayed posters alongside other everyday printed matter, such as business stationery, brochures, and invitations, and, much like the public display of posters, it allowed the material to be circulated in an accessible way that brought the artists' work beyond the confines of avant-garde journals. At the same time, its positioning within museums and other gallery spaces implied the artistic value of everyday printed matter to visitors, opening up a platform for dialogue on its function, as well as broader debates surrounding experimental typography and design.

Kurt Schwitters's typographical work

Schwitters might be best known today for his montages, love/hate relationship with the Dadaists, and the one-man, transdisciplinary art movement he founded, *Merz*, yet the artist worked across many disciplines, including sculpture, painting, poetry, prose, opera, typography, and graphic design. Indeed, Schwitters's creativity spanned, and even challenged, the boundaries of various art forms. Letters, letterforms, and typography are no exception to this, the aesthetics of text being one of the most important parts of his oeuvre. In his montages, Schwitters often includes words that become focal points, in part, due to their striking typography. Indeed, typographical experimentation was one of the hallmarks of his avant-garde journal, *Merz*, in which he played with the size, position, and orientation of the text.[12] In this sense, his later posters can be seen to be an evolution of this practice, and a direct application of the numerous theoretical treatises he wrote on the subject in the 1920s, almost all of which were also published in avant-garde journals.[13]

It is most certainly this practical and theoretical work that led Schwitters to design his most experimental script, the *Systemschrift* [systematic script]. In a letter to his wife dated August 1927, Schwitters opens with news of this project, still in its preliminary phase. Displaying an almost childlike eagerness, he wastes no time with pleasantries, instead jumping straight to the new script in the opening line:

> My dearest!
> I am working on a script. I learnt a lot by visiting the Bauer Type Foundry, e.g. that one must have a pleasant substitute for the difficult letters.[14] [. . .] My basic alphabet is square. Vowels and consonants. Then, for the vowels I also have added a round series. That is an opto-phonetic moment.[15]

The script Schwitters describes is the *Systemschrift* – a series of opto-phonetic scripts, each based on a complex system, that was borne out of a drive towards rationalization and a desire for a modern and efficient way of writing.[16] And so, after years of experimenting with typography across many disciplines, Schwitters turned his thoughts to what is at the very foundation of typography: the alphabet.

At first glance, it may seem that Schwitters was in tune with the work of his colleagues who were also developing new rationalized typefaces, and in part, that was the case. Yet what he was proposing with the *Systemschrift* was not just a new *typeface*, but a new *way of writing*, and was therefore radically different to his design work as a New Typographer. Moving into the fields of linguistics and the philosophy of language, Schwitters aimed to create an entirely new means of rationalized written communication that he felt best suited for the modern age. With this typographical and linguistic project, Schwitters was in fact engaging with much broader questions of internationalism, rationalization, and modes of perception – all of which were circulating at the time, and were starting to have a visible impact on the printed page. A reading of Schwitters's script, and in particular its use in everyday printed matter, therefore provides a useful case study for identifying and exploring these more general trends, as well as drawing attention to the broader implications his work had for the aims of the avant-garde.

Never produced as a typeface, Schwitters's *Systemschrift* remained a theoretical exercise, with the exception of the few published examples examined in this chapter. Apart from a short letter exchange with the Berlin typographer Walter Borgius, the primary source that provides insight into Schwitters's intentions is the article "Ideas for the Adoption of a Systematic Script," and was the only text Schwitters published on his script. Spread across five pages in the avant-garde journal *International Revue i10* in 1927, the article provides a theoretical overview alongside several illustrations.[17] The first of these illustrations presents the reader with six different versions of Schwitters's script under the title "New Plastic Systematic Script" (see Figure 8.1). Differentiated by the letters a–f, lines function as brackets, running down the right-hand side of the page to separate the different

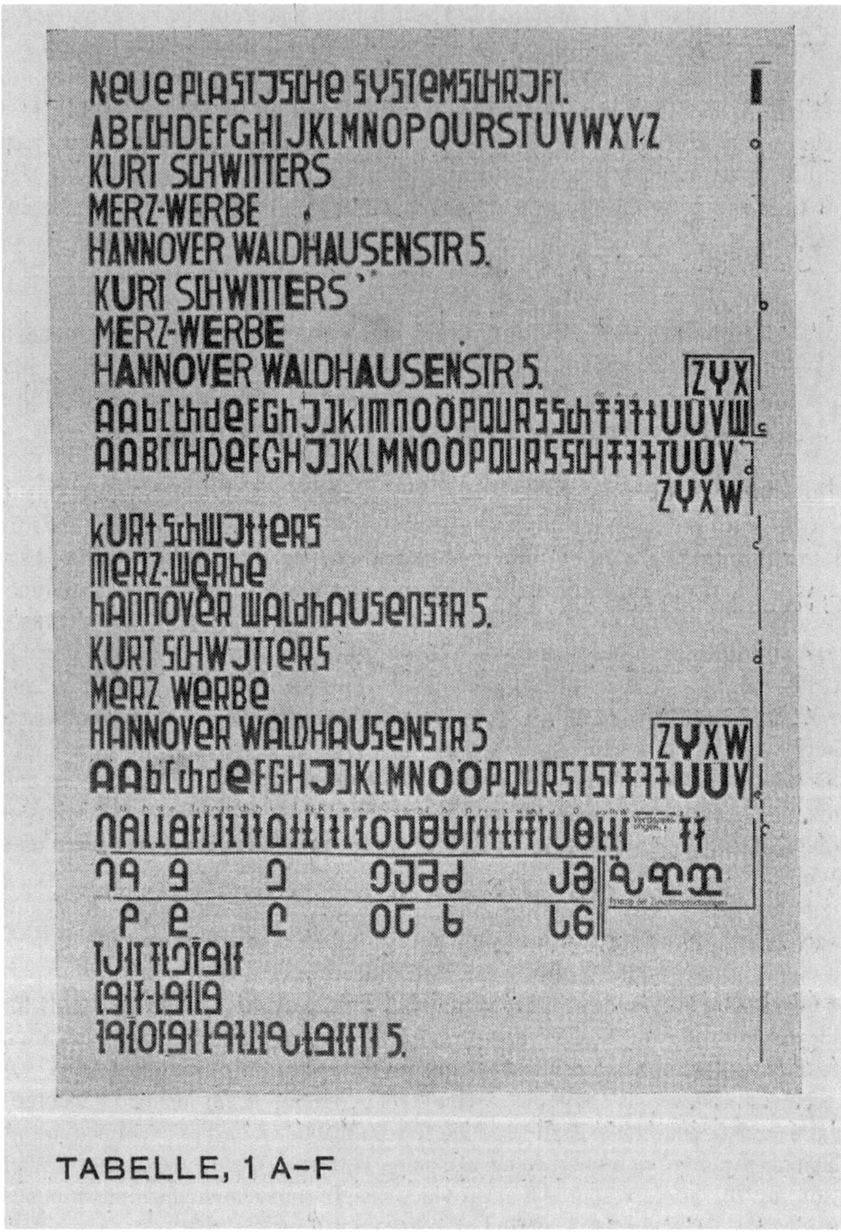

Figure 8.1 Kurt Schwitters, "Tabelle 1, a–f."

Reproduced in Kurt Schwitters, "Anregungen zur Erlangung einer Systemschrift," in *Internationale Revue i 10*, Volume 8/9 (1927), 312.

Source: Photo courtesy of Sprengel Museum Hannover, Herling/Herling/Werner.

prototypes. Alongside each of the lettered versions Schwitters provides a practical example of the script – either by printing his name and address or by transcribing the alphabet. Version "a" very closely resembles other sans-serif typefaces that would have been familiar to the contemporary reader. As the reader's eye moves further down the page, however, not only do the scripts become increasingly unfamiliar, but they also start to include new letters (as in versions "c" and "d"), as well as emphasize vowels (noticeable in script "b," but particularly pronounced by script "e"). The culmination is version "f" which renders both the alphabet and Schwitters's name and address illegible.

"For the purposes of orientation, under a–f are 6 different alphabets," Schwitters writes in a personal letter to his patron Katherine Dreier.[18] "The last [alphabet] is international and very precise, but before being able to read it, it must first be introduced and learnt. c, d, e could already be considered for use."[19] When Schwitters writes of the *Systemschrift*, he does not allude to one typeface or script, as the singular form of the word would have us believe, but rather he refers to an ideal of a script. This ideal existed not just as one, but rather a spectrum of scripts, ranging from the easily recognizable "a" to the radically indecipherable "f."

In explaining the method behind his design of the *Systemschrift*, Schwitters writes, "I have invented an instrument, by means of which, the language can be sorted into three categories. [. . .] The sounds are then ordered according to these categories."[20] The "instrument" to which Schwitters here refers, is a set of tables divided into three columns. After identifying the type and length of each sound in the spoken language, as well as the exact position of the tongue within the mouth, the sound is entered into the table and attributed a corresponding sign. The result is a script that can be used to notate any language (a sincere goal of Schwitters, to be discussed in more detail below), containing so much information, that, once learned, could enable someone who has never before spoken the language to master the pronunciation just by reading the script.

Schwitters produced two posters that use the *Systemschrift*. Both were commissioned pieces of graphic design – the first for the Fitelberg Music Festival (Figure 8.2), and the second for Opel-Tag (Figure 8.3), a day-long celebration of the car manufacturer, Opel. There is much that links these works: both are primarily text based, were produced in 1927, bear not only Schwitters's name as signature, but also the label *Systemschrift*. In addition, both posters were included as illustrations in "Ideas for the Adoption of a Systematic Script."[21]

Although Schwitters indicates that both of these posters use the *Systemschrift*, at first glance the typefaces do not appear to be the same. A closer look reveals that Schwitters uses a different version of the script for each one. The Opel-Tag poster features the less abstracted version "b," which is relatively simple to read and therefore does not immediately strike the viewer as something entirely novel. As the eye moves down the page, differences to the conventional Latin script start to emerge – the bold middle section draws attention to the "J"-like form of the "i" in the text "Prämierung der am schönsten dekorierten Wagen," [award for the best decorated car] as well as in "Eintritt" [entry], found at the bottom

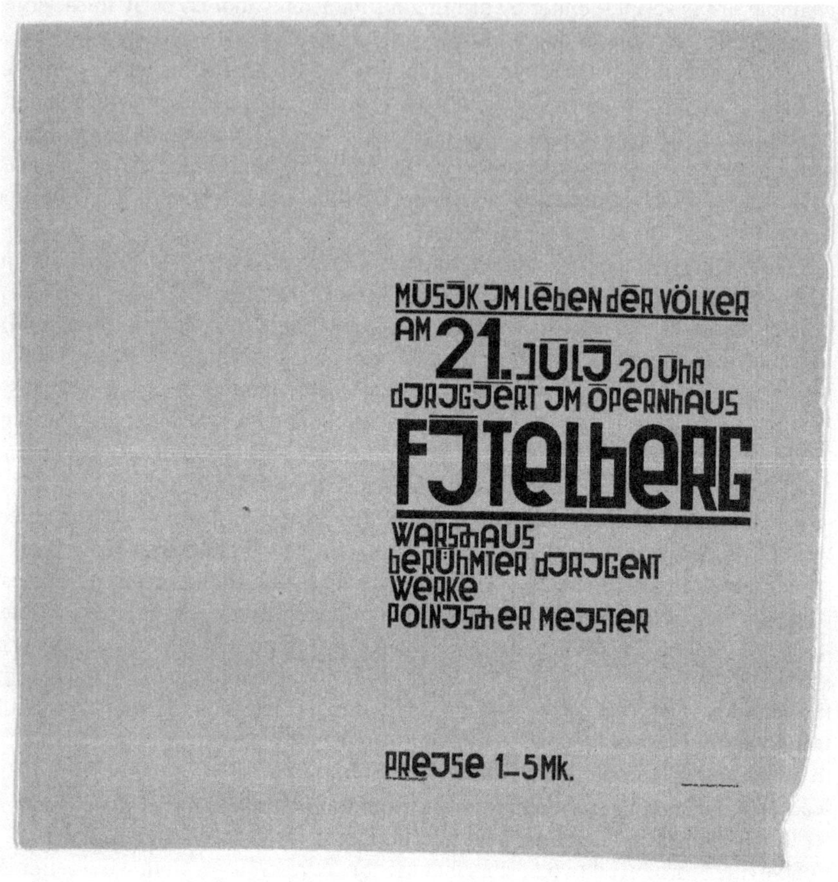

Figure 8.2 Kurt Schwitters: Fitelberg, 1927. Letterpress, 5 7/16 × 5 1/2' (13.8 × 14 cm), Jan Tschichold Collection, Gift of Philip Johnson.

Source: © 2018. Digital image, The Museum of Modern Art, New York/Scala, Florence.

of the page. By contrast, the Fitelberg poster uses something closer to version "e" of the *Systemschrift*, which is furthest down the scale of readability before the essentially illegible "f." Here, the vowels are stressed, not only through their bolder, rounder form, but also by the lines that are placed above longer vowel sounds. As such, it is immediately clear to the reader that this is a new, experimental typeface.

In addition to the typography, there are other significant aesthetic differences between the two designs. The poster advertising the Opel-Tag draws heavily on

Figure 8.3 Kurt Schwitters, poster for Opel-Tag, Frankfurt (1927). Lithograph, 33 1/2 × 23 3/4 inches

Source: Photo courtesy of the Merrill C. Berman Collection.

many New Typography motifs: bold black text, geometric shapes, lines, vertical text, and the stylized arrangement of unused "negative" space. While this remains very much a text-based poster, it acquires an almost geometric form through its composition. The Fitelberg poster, on the other hand, relies much less on any markers of New Typography (although some of these are present), and instead the form is dominated by the unique composition of the individual letters. While the Opel-Tag poster highlights the *Systemschrift*'s common foundation with New Typography, the closer the script becomes to Schwitters's ideal version "f," the more it starts to move towards a different aesthetic, which, rather than emphasizing the composition of the page, instead draws attention to the letters.

There is little information on how either of these commissions came about, what the exact nature of the circulation of these posters was, or indeed how they were received; the events they advertise, however, offer perhaps a few clues. Both the Fitelberg music festival and the Opel-Tag took place in Frankfurt just three days apart and so it is probable that these posters were circulated within the city at the same time. Furthermore, given that the Opel-Tag poster was commissioned by a large, modern company to advertise a display of cars parading through the city, it is likely that this poster in particular would have had a large distribution and be seen on billboards throughout the streets of Frankfurt in the weeks leading up to the event. Both these posters were therefore not only modern in design, but also created for use within the contemporary city, the modern urban subject their imagined audience.

Systemschrift as a modern script

"We have a wealth of typefaces, but all of them are historical, none is systematic," writes Schwitters in "Ideas for the Adoption of a Systematic Script."[22] For Schwitters, the *Systemschrift* was not just a matter of creating something new for the sake of experimentation or novelty. Rather, he was determined to move away from a "historical" script and address what he considered a glaringly inconsistent logic between the rationalized, modern temporality of the 1920s and the antiquated temporality of commonly used inherited scripts. In order to underscore the paradox that he perceived between the conditions of modern life, and the proliferation of text in antiquated fonts, Schwitters employs a vivid image: "It is almost inexplicable," he writes, "that the same people, who today no longer want to ride in the most elegant horse-drawn carriages, use a script that originates in the Middle Ages or antiquity."[23] Here, Schwitters highlights two things simultaneously: the absurdity of clinging on to a medieval script in a society that increasingly privileges speed and functionality over ornamentation and the fragmented and often uneven process of modernization. Although Schwitters in fact rarely employs the word "modern" himself, what he ultimately aims for is a *modern* script, in the sense that the word is associated with the times; that is, in pursuit of ideals of internationalism, rationalization, and the ability to be read at speed.

Principles of internationalism were gaining traction across the field of typography at the time. Herbert Bayer, for example, placed emphasis on the international potential of a new typeface in his 1926 article "Toward a New Alphabet," concluding that international communication could be furthered through the introduction of a *Weltschrift* [world script].[24] While there is no evidence that Bayer ever worked on developing such a *Weltschrift*, in contrast to the traditional German *Fraktur* script, he aimed for his typographic designs to be as clear as possible, facilitating comprehension for non-German readers. Indeed, the name of his best-known typeface, *Universal* – a geometric sans-serif typeface designed in 1925 and composed of lowercase letters only – gestures towards this idea.[25] Meanwhile in Vienna, political economist and philosopher Otto Neurath was also working towards internationalizing language. He writes, "the desire for an international language is an old one, and it is more than ever in men's minds at this time of international connections in business and science."[26] In contrast to Bayer's simplification of the Latin alphabet, Neurath designed a pictorial one, the ISOTYPE (International System of Typographic Picture Education). ISOTYPE has no element of script in it, and is instead made up of simple, repeatable symbols that were used to convey statistics and other numerical information.[27]

A newspaper clipping, cut out and pasted into one of Schwitters's many working notebooks from the mid-1920s, shows the artist's clear interest in internationalism. Taken from a short article by Heinrich Preus entitled "Alongside a World Language, A World Script," the excerpt bears Schwitters's careful pencil marks that underline the following: "We need a world script, we must develop it in such a way, so that each individual sound in any language spoken on earth, is transcribed everywhere in the same way."[28] In contrast to the more pan-European approach to internationalism that Bayer and other New Typographers took, by working primarily on modifying the Latin script, Preus points towards a more inclusive internationalism that takes all languages into account. It is this understanding of concept that Schwitters was striving for with his *Systemschrift*. While he only worked on German sounds, Schwitters saw his work as a single contribution towards a much larger process – an aspect he draws attention to in the opening lines of his article on the script: "A systematic script is only one part of a larger complex problem, which includes a systematic language and systematic thinking. But these are goals for much further in the future."[29] Schwitters knew that it was not possible for himself alone to create what he was ultimately aiming for. In a letter to Borgius he lamented his lack of languages for the task, writing, "I only know German, French, English and Dutch. I also have a little Spanish, Italian, Danish, and Czech."[30] What Schwitters envisaged was not just a European version of internationalism, but a much more radical one, bringing together sounds from all world languages in his *Systemschrift*.

Furthermore, the direction that Schwitters took to achieve a sense of internationalism is very different to that of Bayer and the other New Typographers. While Bayer's typeface *Universal* was based on the principle of simplification of the letterform to render it international, Schwitters aimed to incorporate as much information into each letter as possible. Describing the ideal version "f" of the *Systemschrift*, Schwitters notes that it is "international and very precise."[31] Rather than reducing the typeface to a clean geometric form, free from ornament,

Schwitters changes its form, creating a new, more complex one. Yet the very fact that Schwitters could not use version "f" for either the Fitelberg or Opel-Tag posters highlights the limits of the international reach of the *Systemschrift*. It would have been impossible for him to use it for his advertising commissions, since only Schwitters could read the script, ironically rendering it the very opposite of international. In order for it to become a useable script – and ultimately a truly international one that would reach beyond Europe, or even Frankfurt – it would first have to be understood, and the only way that could happen would be through a process of rationalization.

Rationalization was one of the most important markers of modernity for Schwitters. The emphasis he puts on it in relation to typography is particularly evident in a short text entitled "typography and orthography: lowercase," in which he writes, "Many things move in parallel in the development of our time. The demands of the new era are expressed across many disciplines in the same way." Schwitters continues by stating that there should be "strict laws and logical consistency rather than arbitrariness."[32] For Schwitters, rationalization is the very essence of the time in which he was living and working. In the preceding years there had been several attempts at rationalizing language, but none focused on the transcription of language or the letterform itself.[33] Rationalization was, however, very much in the air of the Weimar Republic. The year 1921 saw the founding of the German State Board for Economy in Industry and Trade, which, according to Mary Nolan "established over two hundred subcommittees to study specific aspects of rationalization."[34] The result was a large restructuring of industry within Germany.[35]

There were two primary ways in which this manifested itself in the printing world: the rationalization of paper size and a debate surrounding the use of uppercase and lowercase letters.[36] In 1922 the *Normenauschuß der deutschen Industrie* [German Industry Standardization Committee] gave birth to the standardized paper size A4, as well as the standardization of envelopes, envelope windows, paper margins, business cards, postcards, logos, and placards. The New Typographers were early adopters of these new formats; the Bauhaus, for example, used DIN formatting for their publicity materials and syllabi, and the school's desks, designed by Vera Meyer-Waldeck, included a drawer specifically for A4 paper.[37] In addition, Jan Tschichold includes a whole section on the value and formatting of standardized paper sizes in his text, *The New Typography: A Handbook for Modern Designers*. Indeed, when publications of any kind – posters, stationery, books, journals – were printed on standardized paper, it was a statement of modernity that often complemented the contents and modern design.

This was equally true for the use of lowercase letters. Many of the New Typographers argued that time, effort, and money could be saved by a process of rationalization by which text would only be written, set, and typed in lowercase. In addition to the arguments set forth in short texts published in journals, reactions to the debate also appeared on other forms of printed matter.[38] Bayer, now famously, states his opinion rather simply with an epigram added to all Bauhaus stationery: "we write everything in lowercase, so as to save time."[39] Similarly, in an exhibition catalogue designed by Schwitters (which I will return to later in this chapter), the following note is included on the opening page: "at the request of professor dr. w. gropius, berlin [. . .] the catalogue has been set in lowercase."[40] It was therefore the printed page – posters, exhibition catalogues, and even stationery – that became the site of theoretical debate, as well as a model of practical application of rationalization.

In his article on the *Systemschrift*, Schwitters draws particular attention to the way in which many other aspects of writing, printing, and textual communication had become systematically normalized, while script itself had remained essentially the same for centuries. Linotype machines had, for example, mechanized the process of typesetting since the end of the nineteenth century, and typewriters were mechanizing the production of text that would have otherwise been handwritten. Meanwhile, the artist felt that the transcription of language had not progressed in anyway, while the world around it had changed drastically. Text was now read in a very different way – it must be read at speed from cars, buses, and bicycles – and therefore needed to be produced in a more standardized manner.

Central to Schwitters's concern with developing a modern script are the implications that modernity and technology had for the reading process. In a letter to his wife, Schwitters poses the question: "What should I call it? A good name is more important than anything else. There are three keywords that are equally important: 'opto-phonetic, communication script [*Verkehrsschrift*], dynamic.' Perhaps you will find a suitable expression."[41] The term *Verkehrsschrift* [communication script] lends us an interesting insight into Schwitters's concept and vision for his script. While it is the name given to a particular form of German shorthand, there is an additional, more common usage of the word *Verkehr*: to mean traffic. Schwitters therefore plays with the notion of communication while also referring to transportation in modern urban life, which is reinforced through the adjective *dynamic*. "It will be so uncomplicated in capital letters," he writes to his wife, "that it will be able to be read quickly, which is absolutely necessary in the rush of traffic."[42] The *Systemschrift* was therefore designed for use within the space of the modern city to facilitate reading at speed.

While Schwitters's theory of the script is firmly rooted in the ideals of internationalism and rationalization, it is exactly *this* that prevents him from achieving his goal of creating a script that can be read at speed. On the one hand, he constructs a very complex system of writing that takes into account *every* sound used in language. Yet as a consequence, Schwitters significantly increases the number of signs used to transcribe German. On the other hand, the whole premise of his short article, "typography and orthography: writing in lowercase," is that systemization calls for the use of small letters in an effort to *reduce* the number of letters. He writes: "if lowercase writing were introduced across the board, the child would only have to learn half the letters, the typesetter could work faster, the printer would only have to buy half the letters, one could type faster on a typewriter, the design of the typewriter could be simplified, etc."[43] There is a distinct tension, therefore, between Schwitters' approach to the rationalization of the existing script and the creation of a new one. Founding partners of *Post Typography* Bruce Willen and Nolen Strals contend that "creating lettering or type is a tug-of-war between the ideal and the practical – the system's concept versus its functionality," and in the case of Schwitters, this plays out on the printed page.[44]

To consider the success of the *Systemschrift* as a modern script, it is helpful to return to the two existing examples in print. Using a less complex, and therefore less rationalized, version of the *Systemschrift*, the Opel-Tag is more easily read by viewers who are not familiar with the script. Yet it is the role of the graphic design of the poster itself that renders parts of it easier to read than others – those sections being the large text "Opel-Tag" and that of the date "24 Juli" [July 24], framed by ample space, as well as the underlining of "freilos" [free raffle ticket] underneath. The typeface, on the other hand, offers the viewer a confusion of round letters (vowels) and angular letters (consonants), that seem to complicate the unified layout of the page. The Fitelberg poster, which is closest to Schwitters's ideal script, magnifies the effect found in the Opel-Tag one. In addition to the distinct differences between vowels and consonants, the changes in letterform and inconsistent thickness of lines break up not just the assumed homogeneity of the text, but also the reader's flow. Indeed, these changes, made in the spirit of rationalization and for the purpose of ease of reading, seem to create the effect of ornamentation – the very thing Schwitters criticizes in his article on the *Systemschrift*. It is ironic, therefore, that the tables of prototypes he sketches reveal six different scripts which become progressively more ornamental. Schwitters's *Systemschrift* posters are striking, but not for their ease of reading, and especially not at speed – it is the new form that catches the eye, yet the script itself slows down the process of reading and therefore also the speed of comprehension.

Beyond the *Systemschrift*: Schwitters and futura

Despite Schwitters's concentrated period of activity on the *Systemschrift*, it is not mentioned in any of his essays, articles, or letters after 1927. The only reference is found in a letter from his wife, Helma Schwitters, to Bodo Rasch in April 1933, in which she replies to the architect's request for samples of the script. "I would send you the photos you wish of the *Systemschrift*, and would have done so long ago, but it is not possible," she writes, continuing, "the progressive work of my husband, must be carried out in secret at the moment."[45] Penned just a few months after the Nazis came to power, Schwitters was obliged to retreat from publicly engaging with any work on his experimental script.

Yet the absence of mention in other texts indicates that Schwitters had stopped working on the *Systemschrift* even before 1933, despite focusing more than ever on typography as the official graphic designer for Hannover city council from 1929 to 1934.[46] In this capacity, Schwitters's work had a far reach, designing more than a hundred different letterheads for use within various departments and institutions governed by the city. In addition to these, Schwitters designed "envelopes, forms and brochures, tickets, posters, and newspaper advertisements" each of which had print runs between 1,000 and 10,000 and were used in schools, hospitals, and offices alike.[47] The publication of bureaucratic forms and headed notepaper, while perhaps seemingly mundane, was, however, significant. Tschichold places particular emphasis on this form of printed matter, noting that the letterhead is "the most important instrument of business communication."[48] The designing and publication of such a document therefore offered Schwitters, and many of his contemporaries, a platform to achieve what Peter Bürger identifies as one of the main aims of the historical avant-garde: "the sublation of art in the praxis of life."[49] Letters bearing Schwitters's designs for Hannover council were sent and received on a daily basis, leaving institutions and entering the homes and daily lives of *all* residents of Hannover. Like the Opel-Tag and Fitelberg posters, then, Schwitters graphic design work for the council was functional ephemeral art – art that was not intended to be viewed in a museum, but rather to be touched, written upon, and otherwise utilized towards practical ends.

Instead of using one of his own typefaces, however, for the majority of the Hannover council stationery Schwitters opted for futura, which he described as clear, plain, easily readable, and precise – the very qualities he had been striving for with the *Systemschrift*.[50] Futura was created by book designer Paul Renner, released in 1927 (the same year Schwitters worked on the *Systemschrift*), and would very quickly prompt major change in the publishing world. Clean, geometric, and sans-serif, futura marked a typographical watershed.[51] It was a direct response to the *Antiqua-Fraktur* debate and was widely embraced by the New Typographers. Indeed, Schwitters's use of it in his designs for the Hannover council helped make it a more commonplace typeface, since, as Heine notes, "many printers [in Hannover], especially smaller ones, did not possess Futura, and were forced to acquire it in order to remain competitive."[52] Once in the printer's shop, it could therefore be used more readily by other clients.

Soon after the Nazis came to power, however, Schwitters was forced to change his designs for the council from futura to *Fraktur* when a memorandum on June 14, 1933, called for the end of the sans-serif typeface on all official documents.[53] This was the fate of the work of many of the New Typographers; as a marker of Weimar modernity, their typefaces and graphic designs were deemed "un-German" by the Nazis, with *Fraktur* being considered the most "fitting" script for German texts.[54]

Legacy of Schwitters's designs

Since the 1950s, futura and similar typefaces have become commonplace, but what about Schwitters's aim for an entirely rationalized way of writing? As late as 1980 Bodo Rasch and Herbert Bayer were still working with concepts of a new form of written communication.[55] In a letter to Rasch, Bayer writes,

> I thank you for sending your lengthy manuscript on your thoughts about writing, and the new alphabet. I am much impressed by the thoroughness with which you have approached this complex problem, and I am indeed happy that you have taken up the idea of an improved writing method. as to my disappointment, nothing of any value has come from the side of the graphic designers. they are all interested only in the improvement of writing as a tool.

> [. . .]
> I have never really known kurt schwitters' ideas about an opto-phonetic alphabet, but I am convinced that it would be worthwhile to consider some of his thoughts.[56]

In expressing his frustration that, in the half-century that had passed since the 1930s, no progress had been made, Bayer directs Rasch back to Schwitters, as a point of reference for considering a new type of script. Several things can be gleaned from this letter; first, Schwitters's *Systemschrift* was widely known in avant-garde circles, and significant enough to be spoken about over fifty years later. Second, while there has been a continued trend of geometric, sans-serif typefaces first introduced by the New Typographers, no one has since taken up the project of creating a new, rationalized, and truly international opto-phonetic script.

Although the rationalized ideal of Schwitters's version "f" of the *Systemschrift* has lain dormant since 1927, other versions of his script have not completely disappeared. On the contrary, there have been three recent examples of its use. The first of these appeared in 1997 – exactly seventy years after Schwitters made preliminary sketches of it – with the release of the typeface *Architype Schwitters* by The Foundry, run by two British typographers, David Quay and Freda Sack.[57] This typeface is a faithful combination of versions "c" through "e" of Schwitters's *Systemschrift*, and since it is available digitally has the potential for a much more global application of the script than Schwitters ever achieved in his lifetime.

The second example can be found by logging onto the International Online Bibliography of Dada at the University of Iowa.[58] A banner across the top reads "International Dada Archive," written in a rendering of Schwitters's *Systemschrift*. This is not the same typeface created by The Foundry and instead is much closer to the version Schwitters used for the Fitelberg poster, as evidenced by the stresses above certain vowels. The context in which this script is used draws attention to its experimental qualities and serves two functions. First, on a basic level, it provides the reader with textual information; and second – and more important – Schwitters's script comes to aesthetically represent the archive.[59] While this recent example gives the *Systemschrift* a certain new life by using it in a way Schwitters never did, at the same time, in symbolizing an archive, it continually points back to the temporal and geographical context from which it was born – the interwar years in Germany.

The third and final example is not strictly one of the *Systemschrift*, but rather another experimental typeface designed by the artist that shares many of characteristics of versions "a" through "e." In 1929, hot on the heels of his appointment as graphic designer for Hannover, Schwitters won the contract to create stationery and advertisements for the *Dammerstock-Siedlung* and the exhibition that would mark its opening. Set in the outskirts of Karlsruhe, the Dammerstock-Siedlung is a housing estate designed by a handful of invited architects who were working in the style of *Neues Bauen* [New Building] and led by Bauhaus founder Walter Gropius. As a modern rationalized style of architecture, prizing functionality and affordability, *Neues Bauen* was not just contemporaneous to New Typography, but also had a certain amount of overlap with its ideals.[60]

In addition to creating a letterhead, envelope, and logo for the Dammerstock-Siedlung, Schwitters's main responsibility was to design the housing estate's exhibition catalogue and poster. Brigitte Franzen notes that there was a major advertising campaign surrounding the exhibition, which drew 50,000 visitors over the course of the month it was open to the public.[61] Similar to his work for the Hannover council, this project allowed Schwitters's typographical and graphic designs to reach a broad public and afforded him the opportunity to merge art with everyday life – both of which were also central premises of *Neues Bauen*. Indeed, the sheer scale of this advertising project meant that Schwitters's work was displayed throughout the city of Karlsruhe and beyond in several different formats.

The Dammerstock-Siedlung logo, comprising a black block in the shape of the housing development, with a large lowercase "d" emerging from it to form the first letter of "Dammerstock," appears on the poster as well as the exhibition catalogue. On both, the following text, all in uppercase, sets forth the title, "Ausstellung Karlsruhe Dammerstock-Siedlung: die Gebrauchswohnung" [Karlsruhe's Dammerstock Estate Exhibition: The Utility Apartment], which is set in a typeface designed by Schwitters. Although not exactly the same as any version of the *Systemschrift*, the consonants strongly resemble those of version "d" with only the squareness of the vowels distinguishing this script from Schwitters's 1927 prototypes. It is not just the similarity between these two scripts that suggests the Dammerstock script is related to the *Systemschrift* – the A4-format of the catalogue, as well as the fact the text on the pages inside the catalogue is set in a lowercase, sans-serif typeface (Renner's futura) all point towards an ideal of rationalization. Even if not considered by Schwitters as a later version of his *Systemschrift*, his designs for the Dammerstock project not only borrow aesthetically from his earlier work, but are also very much ideologically aligned with it.

If 1927 marked a high point for Schwitters's focus on rationalizing writing, the Dammerstock script marks a retreat back from his radical version "f," and a step towards a script that would have been easily legible to the contemporary reader. It was perhaps in part the immensity of this particular project that had the potential to reach so many people in their everyday lives that prompted him to instead focus on adapting the Latin script, rather than pushing for a new way of writing.

Just as Schwitters conceived of the *Systemschrift* as a script for the city, the Dammerstock script became one that not only represented an area of the city, but was seen throughout Karlsruhe in 1929. A photograph from that time depicts the scene outside the train station: a free-standing sign with the title of the exhibition, inscribed in the same typeface as that used on the poster and catalogue, points in the direction of the housing development (Figure 8.4). To the right, the exhibition poster is positioned halfway up, with the word *Omnibus* printed in the Dammerstock typeface. Unlike the other signs visible in the photograph, the text on the sign is printed onto glass, giving the impression that Schwitters's text is superimposed directly onto the street beyond it. As a result, his script appears to physically meld with the city. However, just like the temporary nature of the exhibition, the script on the sign and posters across the city soon disappeared from the streets.

Figure 8.4 Dammerstock-Siedlung Exhibition, 1929. A sign points toward the Dammerstock-Siedlung exhibition featuring a typeface created by Schwitters for the project. To the right, a poster advertising the exhibition, designed by Schwitters, is also visible.

Photo by Adolf Supper, 1929. "Eröffnung der Dammerstocksiedlung – Hinweisschild 'Ausstellung-Dammerstock-Siedlung, Die Gebrauchswohnung.'" Stadtarchiv Karlsruhe, 8/Alben 398 / 2.

Source: Photo courtesy of Stadtarchiv Karlsruhe.

In 2006, the discovery and refurbishment of one of two original ticket counters used at the entrance to the 1929 exhibition brought Schwitters's Dammerstock script back to the streets of Karlsruhe.[62] This ticket counter now sits to the southeast entrance to the housing development, beside a new structure that serves as an information pavilion, with a roof straddling the two (Figure 8.5). Like the 1929 sign, the information pavilion is made out of glass and bears the text, "Dammerstock Muster-Siedlung des neuen Bauens" [Dammerstock Model Estate of New Building], in the script used on Schwitters's exhibition catalogue and poster, with the logo positioned behind it. The same script also makes an appearance on the original ticket counter, on which raised wooden letters form the text "Eintritt 50 Pfennig" [entrance 50 pfennig]. Consequently, Schwitters's script has been reinscribed onto the city with two consequences. First, the characteristic form of the script has come to be a symbol for the Dammerstock-Siedlung. Second, it is also used in the way in which Schwitters always intended it – as a means of conveying information in a fast-paced, modern, urban environment. Indeed, directly in front of the ticket counter and information pavilion today, just a small strip of grass separates it from the commuter railway tracks, with the pavilion located exactly between two stops, and beyond that, lies

Figure 8.5 Infopavillon at the entrance to the Dammerstock-Siedlung (2017).
Source: Photo courtesy of the author.

a busy main road. Schwitters's text is therefore seen each day by car and train passengers who catch a glance as they travel past.

While this use might seem to be in line with Schwitters's goal of his experimental typography being used in the city, much like the International Dada Archive's use of his *Systemschrift*, the musealization of the Dammerstock script in the information pavilion also implies its historicization. Although present in the streets of Karlsruhe today, it is no longer used to point towards an innovative new housing estate or modern temporary exhibition, but is rather used to evoke an aesthetic link to the past.

The legacy of Schwitters's experimental typography is far from large and is most likely restricted to the three examples discussed above, each of which closely resemble the existing Latin script. As such, they remain estranged from the artist's ideal of an entirely new, international, and rationalized way of writing – an idea that has seemingly not evolved since the drafts he created in 1927. Although Schwitters's most radical scripts were not entirely successful, the project was significant, extending beyond his lifetime.

Most important, however, Schwitters's graphic design work offered him a diverse range of publishing platforms – be it poster, letterhead, or magazine – to enter into dialogue with other New Typographers and theorists. As these examples have shown, the nature of this graphic design work and the ways in which it was circulated allowed him to successfully realize certain aspects of the avant-garde project of merging art and life, therefore repositioning this aspect of his work as pivotal to his overall creative aims. Such a reading therefore nudges towards a need to reconsider the role that graphic design and typography – together with the platforms in which they were published – plays in Schwitters's overall oeuvre, and by extension, that of many other avant-garde artists.

Notes

1 This essay is born out a chapter from my dissertation and as such, I would like to thank in particular my dissertation chair Kerstin Barndt for her inestimable feedback on earlier versions of this work.

2 At the heart of this debate lay the question of which typeface was the most suitable for publishing German books and newspapers. At one pole end was *Fraktur* (also known as Gothic), a Blackletter script (a Gothic script commonly used in Europe in the medieval age) that had been Germany's primary typeface since the mid-sixteenth century; at the other was *Antiqua* – a family of Roman scripts that dates back to the late fifteenth century and remains in common use today (Times New Roman, etc.), that replicate handwriting more closely than the ornamental *Fraktur*. While *Fraktur* had evolved into a German script, it shared roots with various Blackletter typefaces that had been common throughout Europe since the twelfth century. In the following centuries, however, they fell out of favor in other western and northern European countries. Great Britain ceased to use it at the beginning of the seventeenth century, in Sweden they changed to Roman script in the eighteenth century, and France followed suit at the beginning of the nineteenth century.

The arguments that governed the *Antiqua-Fraktur* debate fell into two main categories: first, a political and ideological debate that centered on notions of nationalism versus internationalism, and second, the importance of functionality for a script. With regards to the political and ideological debate, the arguments of those in favor of *Fraktur* propagated ideas of nationalism – its supporters claimed that as a German script it was the most suitable for publishing German books and newspapers. Meanwhile, proponents of *Antiqua* saw the international readability of the script as one of its most important advantages.

For more on the *Antiqua-Fraktur* debate see Christopher Burke, "German Hybrid Typefaces 1900–1914," in *Blackletter: Type and National Identity* (New York: Cooper Union, 1998), 32–39; Silvia Hartmann, *Fraktur oder Antiqua: der Schriftstreit von 1881 bis 1941* (Berlin: Peter Lang Publishing, 1998).

3 Herbert Bayer was both a student and professor of typography at the Bauhaus.

László Moholy-Nagy was a professor at the Bauhaus, who wrote several theoretical essays on typography, and incorporated his theoretical statements into his graphic design work.

Like Bayer, Josef Albers was both a student and professor at the Bauhaus, who designed the geometric *Schablonenschrift* [stencil script].

4 The term "sans-serif" is a general one that refers to all typefaces that do not have serifs – the strokes that protrude at the ends of letters. Sans-serif typefaces first emerged during the nineteenth century, but did not become commonplace until the 1920s, when New Typographers paved the way for their more widespread use.

5 While the term today is most commonly associated with Jan Tschichold's seminal text *Die neue Typografie*, published in 1928 [Jan Tschichold, *The New Typography: A Handbook for Modern Designers*, trans. Ruari McLean (Berkeley and Los Angeles: University of California Press, 1995)], it was first used by Moholy-Nagy in 1923 for the catalogue of the first Bauhaus exhibition.

6 See Julia Meer, *Neuer Blick auf die Neue Typographie: Die Rezeption der Avantgarde in der Fachwelt der 1920er Jahre* (Bielefeld: Transcript Verlag, 2015), 28–29.

7 For examples of turn-of-the-century posters, see for instance the work of Czech artist Alphonse Mucha, who studied in Munich, or that of Vienna Secession artist Koloman Moser.

8 Kurt Schwitters, Käte Steinitz, and Theo Van Doesburg, *Die Scheuche: Märchen* (Hannover: Aposs Verlag, 1925).
9 See Tschichold, *The New Typography*.
10 Megan R. Luke, *Kurt Schwitters: Space, Image, Exile* (Chicago: University of Chicago Press, 2014), 38.
Several versions of the *neue Typografie* exhibition were on display in many different cities including Cologne, Wiesbaden, Hannover, Bremen, Hamburg, Dresden, Basel, Rotterdam, and Amsterdam.
11 The founding members were Willi Baumeister, Max Burchartz, Walter Dexel, César Domela, Robert Michel, Kurt Schwitters, Georg Trump, Jan Tschichold, and Friedrich Vordemberge-Gildewart.
12 For an example of this, see Kurt Schwitters, "Merz 1: Holland," in *Das literarische Werk: Prosa 1918–1930*, ed. F. Lach, vol. 5 (Cologne: M. DuMont Schauberg, 1981), 124–132.
13 See, for example, Kurt Schwitters, "Thesen über Typographie," *Merz* 11 (1925); Kurt Schwitters, "Gestaltende Typographie," in *Das literarische Werk: Prosa 1918–1930*, ed. F. Lach, vol. 5 (Cologne: M. DuMont Schauberg, 1981), 311–316; Kurt Schwitters, *Werbe-Gestaltung* (Hannover: Merz, 1928).
14 This is the same type foundry that produced Paul Renner's futura typeface (referred to later in this chapter), which was released the same year.
15 Schwitters to Helma Schwitters, Bad Ems, 14 August 1927, in Kurt Schwitters, *Wir spielen, bis uns der Tod abholt: Briefe aus 5 Jahrzehnten*, ed. Ernst Nündel (Berlin: Ullstein, 1986), 127.
16 Schwitters was not the only avant-garde artist working with the idea of opto-phonetics. Raoul Hausmann had been working with the concept in relation to his visual poetry, which was based in experimental typography. In the journal *MA* he writes in 1922, "If there is an appropriate technical structure, the optophone has the power, or rather the ability to display the audio equivalent of each optical phenomenon." [Reproduced in English translation in Peter Weibel, ed., *Beyond Art: A Third Culture: A Comparative Study in Cultures, Art and Science in 20th Century Austria and Hungary* (Berlin: Springer Science & Business Media, 2005), 83]. El Lissitzky pushed for a similar idea in his manifesto "Topography of Typography," published in issue 4 of *Merz*, with his fourth point, which reads: "Economy of expression – optics not phonetics" [El Lissitzky, "Topographie der Typographie," *Merz*, no. 4 (July 1923): 47].
17 Original title: "Anregungen zur Erlangung einer Systemschrift." An abridged, one-page version of this article appeared a year later in *Der Sturm*.
18 Katherine Dreier was an American artist, patron of the arts, and cofounder of *The Société Anonyme*, together with Man Ray and Marcel Duchamp.
19 Schwitters, to Katherine S. Dreier, Eppstein, June 27, 1927, in Kurt Schwitters, *Wir spielen, bis uns der Tod abholt: Briefe aus 5 Jahrzehnten*, ed. Ernst Nündel (Berlin: Ullstein, 1986), 127.
20 Schwitters, to Walter Borgius, Bad Ems, July 2, 1927, in Schwitters, *Wir spielen, bis uns der Tod abholt*, 121.
21 Both Schwitters's name and the label *Systemschrift* are found in very small type in the left-hand corner of the Fitelberg poster and to the right of center at the bottom in the Opel-Tag poster.
22 Schwitters, "Anregungen zur Erlangung einer Systemschrift," 274.
23 Ibid.
24 Herbert Bayer, "Versuch einer neuen Schrift," in *Bauhaus: Drucksachen, Typografie, Reklame*, ed. Gerd Fleischmann (Dusseldorf: Ed. Marzona, 1984), 25.
25 Bayer is quick to note that there is already a typeface that is being used in an international way – the *Grotesk* script [see ibid., 26] – one of the three main families of script, together with *Antiqua* and *Fraktur* [see Frank Koschembar, *Grafik für Nicht-Grafiker: ein Rezeptbuch für den sicheren Umgang mit Gestaltung; ein Plädoyer für besseres Design* (Frankfurt am Main: Westend-Verlag, 2005), 18]. *Grotesk* is a more modern

version of *Antiqua*, having been first introduced in the nineteenth century, and is sans-serif. Yet Bayer points out the limits of the script, stating, "It is used internationally, but it is still neither perfect nor uncomplicated, as it has been arbitrarily developed." [Bayer, "Versuch einer neuen Schrift," 26].

26 Otto Neurath, *International Picture Language: The First Rules of Isotype with Isotype Pictures* (London: Kegan Paul, Trench, Trübner & Company, 1936), 16.

Otto Neurath (1882–1945) was a political economist, philosopher of science, and member of the Vienna Circle. Throughout his career Neurath was interested in how to effectively disseminate information to the public, and was director of the *Deutsches Kriegswirtschaftsmuseum*, before founding the *Siedlungsmuseum*, and later the *Gesellschafts- und Wirtschaftsmuseum*. For more on Otto Neurath, see Nancy Cartwright, Jordi Cat, Lola Fleck, and Thomas E. Uebel, *Otto Neurath: Philosophy Between Science and Politics* (Cambridge, UK: Cambridge University Press, 2008).

27 Ellen Lupton notes that each of the individual characters that make up ISOTYPE is "similar to a scientific formula, it is a reduced and conventionalized scheme of direct experience. [. . .] An Isotype character formulates the undifferentiated, nonhierarchical details of the photograph into a concise, repeatable, generalized scheme." [Ellen Lupton, "Reading Isotype," in *Design Discourse: History, Theory, Criticism*, ed. Victor Margolin (Chicago: University of Chicago Press, 1989), 146]. This language was subsequently employed by Neurath in the *Gesellschafts- und Wirtschaftsmuseum*, and although ISOTYPE itself had no words, in the museum it appeared alongside text composed in Renner's futura, perhaps as a result of Jan Tschichold's time there. [See Christopher Burke, *Active Literature: Jan Tschichold and New Typography* (Princeton, NJ: Princeton Architectural Press, 2007), 119].

For more information on ISOTYPE see Christopher Burke, Eric Kindel, Sue Walker, et al., *Isotype: Design and Contexts 1925–1971* (London: Hyphen Press, 2013); Christopher Burke, "Isotype: Representing Social Facts Pictorially," *Information Design Journal* 17, no. 3 (2009): 211–223; Lupton, "Reading Isotype."

28 Heinrich Preus, "Neben eine Weltsprache," in Kurt Schwitters Papers, Kurt Schwitters Archive, Sprengel Museum, Hannover.

29 Schwitters, "Anregungen zur Erlangung einer Systemschrift," 274.

30 Schwitters, to Walter Borgius, Eppstein, July 17, 1927, in Kurt Schwitters, *Wir spielen, bis uns der Tod abholt: Briefe aus 5 Jahrzehnten*, ed. Ernst Nündel (Berlin: Ullstein, 1986), 125.

31 Schwitters, to Katherine S. Dreier, Eppstein, June 27, 1927, 120.

32 Kurt Schwitters, "typographie und orthographie: kleinschrift," in *Das literarische Werk: Prosa 1918–1930*, ed. F. Lach, vol. 5 (Cologne: M. DuMont Schauberg, 1981), 268.

33 One such example is Esperanto, a planned language designed by Ludwik Zaemnhof in the 1880s to help overcome communication issues resulting from the multilingualism of an international world. Although still in use today, it remains very much a minority language spoken primarily by those whose mother tongue is a European language. For more information on Esperanto, its implications, and importance, see Pierre Janton and Humphrey Tonkin, *Esperanto: Language, Literature, and Community* (Albany: SUNY Press, 1993).

It is important to note that Esperanto was not the only international language at this time. Around the same time Zaemnhof was working on Esperanto, the German Roman Catholic priest Johann Martin Schleyer was creating *Volapük*. Schleyer claimed that God had spoken to him in a dream and asked him to create an international language. For more information on *Volapük* see Andrew Drummond, *A Handbook of Volapük* (Edinburgh: Polygon, 2006).

34 The German title for the State Board for Economy in Industry and Trade is *Reichskuratorium für Wirtschaftlichkeit in Industrie und Handwerk* (RKW).

Mary Nolan, *Visions of Modernity: American Business and the Modernization of Germany* (Oxford: Oxford University Press, 1994), 135.

35 Ibid., 133.

36 One of the first proponents of omitting uppercase in favor of lowercase letters was Walter Porstmann, a German engineer and mathematician, who worked on standardization and introduced the DIN (*Deutsche Instituts für Normung*, or German Insitute for Standardization) format paper size (A4 etc.), which continues to be the standard paper size throughout the world, with the exception of North America. His 1920 monograph *Sprache und Schrift* [Language and Script] came to be one of the most influential texts for the New Typographers. In it, he points out "if we count a German text, for every hundred letters we find around five uppercase letters. For five percent of our writing we therefore burden the whole writing process – from learning to application – with double the number of signs." He conclucles by underlining this point in terms of economy, "because of five percent of the letters, we produce a hundred percent increase in the letters."

Walter Porstmann, *Sprache und Schrift*, ed. Richard R. Hinz (Berlin: Verlag des Vereins Deutscher Ingenieure, 1920), 70.

37 Magdalena Droste, *Bauhaus, 1919–1933* (Cologne: Taschen, 2002), 194.

38 See, for example, Schwitters, "typographie und orthographie," 268–269.

39 For an example, see a sheet of headed notepaper reproduced in Gerd Fleischmann, ed., *Bauhaus: Drucksachen Typografie, Reklame* (Düsseldorf: Ed. Marzona, 1984), 117.

40 Walter Gropius, *Dammerstock Katalog: Ausstellung Karlsruhe Dammerstock-Siedlung: Die Gebrauchswohnung* (Karlsruhe: Stadt Karlsruhe, 1929), 1.

41 Schwitters to Helma Schwitters, Bad Ems, August 14, 1927.

42 Ibid.

43 Schwitters, "typographie und orthographie," 269.

44 Bruce Willen and Nolen Strals, *Lettering & Type: Creating Letters and Designing Typefaces* (Princeton, NJ: Princeton Architectural Press, 2009), 21–22.

45 Helma Schwitters to Bodo Rasch, 5 April 1933. Bodo Rasch Papers, D2179, ADK 10–21/82, Werkbund Archiv, Berlin.

46 For more on this aspect of Schwitters work, see Werner Heine, "'Futura' without a Future: Kurt Schwitters' Typography for Hannover Town Council, 1929–1934," trans. Annette Haxton, *Journal of Design History* 7, no. 2 (1994): 127–140.

47 Heine, "'Futura' without a Future," 129, 131.

48 Tschichold, *The New Typography*, 112.

49 Peter Bürger, *Theory of the Avant-Garde* (Manchester: Manchester University Press, 1984), 51.

50 Kurt Schwitters, "Werbe-Gestaltung," in *Das literarische Werk: Prosa 1918–1930*, ed. F. Lach, vol. 5 (Cologne: M. DuMont Schauberg, 1981), 13.

There are some exceptions to this. Heine points out that Schwitters incorporates his own typeface into two programs for the State Theater in Hannover in 1932. Heine, "'Futura' without a Future," 132.

51 For more on the impact futura and Paul Renner had on typography, see Christopher Burke, *Paul Renner: The Art of Typography* (Princeton, NJ: Princeton Architectural Press, 1998).

52 Heine, "'Futura' without a Future," 130.

53 See Ibid., 136.

54 As a result, several New Typographers, including Paul Renner and Jan Tschichold, were forced into exile in Switzerland, with the seat of European typography shifting from Germany to its neighbor from the mid-1930s onward. Consequently, the *International Typographic Style*, now more commonly known as *Swiss Style*, emerged from Switzerland during the 1940s and 1950s. In many respects, this new typographic movement can be seen to be a direct continuation of New Typography. In addition to a shared goal of internationalism, in practice, the Swiss Style typefaces, like those of the New Typographers, were sans-serif and based on geometric grids and asymmetry. See Richard Poulin, *Graphic Design and Architecture, a 20th Century History: A Guide to Type, Image, Symbol, and Visual Storytelling in the Modern World* (Beverly, MA: Rockport Publishers, 2012), 133.

55 Bodo Rasch (1903–1995) was an architect. and together with his brother Heinz Rasch edited one of the most important publications on New Typography, *Gefesselter Blick* [Heinz Rasch and Bodo Rasch, *Gefesselter Blick* (Stuttgart: Wissenschaftlicher Verlag Dr. Zaugg & Co., 1930)].

56 Herbert Bayer to Bodo Rasch, November 19, 1980, Bodo Rasch Papers, D2179, ADK 10–21/82, Werkbund Archiv, Berlin. (This letter was originally composed in English, in all lowercase, with the exception of the "I," which always appears in uppercase.)

57 *Architype Schwitters* is part of a series of typefaces based on those created by experimental typographers and New Typographers during the interwar period. The collection also includes *Architype Albers*, *Architype van Doesburg*, *Architype Bayer*, *Architype Tschichold*, and *Architype Renner*.

58 "International Dada Archive," accessed March 30, 2018, http://search.lib.uiowa.edu/primo_library/libweb/action/search.do?vid=01IOWADADA.

59 It is ironic that a typeface by Schwitters, who throughout his life positioned *Merz* in opposition to the work of the Dadaists, should be chosen to visually represent the International Dada Archive.

60 For more information on the Dammerstock-Siedlung, see Peter Schmit and Brigitte Franzen, *Neues Bauen der 20er Jahre: Gropius, Haesler, Schwitters und die Dammerstocksiedlung in Karlsruhe 1929* (Karlsruhe: Museum für Moderne Angewandte Kunst, 1997).

61 Brigitte Franzen, "Einleitung," in *Neues Bauen der 20er Jahre: Gropius, Haesler, Schwitters und die Dammerstocksiedlung in Karlsruhe 1929*, ed. Brigitte Franzen and Peter Schmitt (Karlsruhe: Museum für Moderne Angewandte Kunst, 1997), 20.

62 The second ticket counter was dismantled after the exhibition. The refurbished one was used as a ticket desk at an outdoor swimming pool in Rappenwört, just outside of Karlsruhe, until its discovery in 2004. Subsequently, the architect firm Rossmann+Partner, whose offices have been in the main *Waschhaus*, situated at the entrance of the Dammerstock since 1974, refurbished it. Today it is part of the Infopavillion with information on the history of the Dammerstock exhibition.

For more information, see "Infopavillion Dammerstock," accessed, April 19, 2018, http://rossmannundpartner.de/projekte/infopavillion-dammerstock/.

Bibliography

Bürger, Peter. *Theory of the Avant-Garde*. Manchester, UK: Manchester University Press, 1984.

Burke, Christopher. *Active Literature: Jan Tschichold and New Typography*. Princeton, NJ: Princeton Architectural Press, 2007.

Burke, Christopher. "German Hybrid Typefaces 1900–1914." In *Blackletter: Type and National Identity*, 32–39. New York: Cooper Union, 1998.

Burke, Christopher. "Isotype: Representing Social Facts Pictorially." *Information Design Journal* 17, no. 3 (2009): 211–223.

Burke, Christopher, Eric Kindel, SueWalker, eds. *Isotype: Design and Contexts 1925–1971*. London: Hyphen Press, 2013.

Cartwright, Nancy, Jordi Cat, Lola Fleck, Thomas E. Uebel, eds. *Otto Neurath: Philosophy between Science and Politics*. Cambridge, UK: Cambridge University Press, 2008.

Droste, Magdalena. *Bauhaus, 1919–1933*. Cologne: Taschen, 2002.

Drummond, Andrew. *A Handbook of Volapük*. Edinburgh: Polygon, 2006.

Fleischmann, Gerd, ed. *Bauhaus: Drucksachen, Typografie, Reklame*. Düsseldorf: Ed. Marzona, 1984.

Gropius, Walter. *Dammerstock Katalog: Ausstellung Karlsruhe Dammerstock-Siedlung: Die Gebrauchswohnung*. Karlsruhe: Stadt Karlsruhe, 1929.

Hartmann, Sylvia. *Fraktur oder Antiqua: der Schriftstreit von 1881 bis 1941*. Berlin: Peter Lang Publishing, 1998.

Heine, Werner. "'Futura' without a Future: Kurt Schwitters' Typography for Hannover Town Council, 1929–1934." Translated by Annee Haxton. *Journal of Design History* 7, no. 2 (1994): 127–140.

Janton, Pierre and Humphrey Tonkin. *Esperanto: Language, Literature, and Community*. Albany: SUNY Press, 1993.

Koschembar, Frank. *Grafik für Nicht-Grafiker: ein Rezeptbuch für den sicheren Umgang mit Gestaltung; ein Plädoyer für besseres Design*. Frankfurt am Main: Westend-Verlag, 2005.

Luke, Megan R. *Kurt Schwitters: Space, Image, Exile*. Chicago: University of Chicago Press, 2014.

Lupton, Ellen. "Reading Isotype." In *Design Discourse: History, Theory, Criticism*, edited by Victor Margolin, 145–156. Chicago: University of Chicago Press, 1989.

Meer, Julia. *Neuer Blick auf die Neue Typographie: Die Rezeption der Avantgarde in der Fachwelt der 1920er Jahre*. Bielefeld: Transcript Verlag, 2015.

Neurath, Otto. *International Picture Language: The First Rules of Isotype with Isotype Pictures*. London: Kegan Paul, Trench, Trübner & Company, 1936.

Nolan, Mary. *Visions of Modernity: American Business and the Modernization of Germany*. Oxford: Oxford University Press, 1994.

Porstmann, Walter. *Sprache und Schrift*. Edited by Richard R. Hinz. Berlin: Verlag des Vereins Deutscher Ingenieure, 1920.

Poulin, Richard. *Graphic Design and Architecture, a 20th Century History: A Guide to Type, Image, Symbol, and Visual Storytelling in the Modern World*. Beverly, MA: Rockport Publishers, 2012.

Rasch, Heinz and Bodo Rasch. *Gefesselter Blick*. Stuttgart: Wissenschaftlicher Verlag Dr. Zaugg & Co., 1930.

Schmitt, Peter and Brigitte Franzen, eds. *Neues Bauen der 20er Jahre: Gropius, Haesler, Schwitters und die Dammerstocksiedlung in Karlsruhe 1929*. Karlsruhe: Museum für Moderne Angewandte Kunst, 1997.

Schwitters, Kurt. *Das literarische Werk: Prosa 1918–1930*. Edited by F. Lach, Vol. 5. Cologne: M. DuMont Schauberg, 1981.

Schwitters, Kurt. "Thesen über Typographie." Edited by Kurt Schwitters and El Lissitzky. *Merz* 11 (1925).

Schwitters, Kurt. *Werbe-Gestaltung*. Hannover: Merz, 1928.

Schwitters, Kurt. *Wir spielen, bis uns der Tod abholt: Briefe aus 5 Jahrzehnten*. Edited by Ernst Nündel. Berlin: Ullstein, 1986.

Schwitters, Kurt, Käte Steinitz, and Theo Van Doesburg. *Die Scheuche x: Märchen*. Hannover: Aposs Verlag, 1925.

Toscano, Aaron. *Marconi's Wireless and the Rhetoric of a New Technology*. Berlin: Springer Science & Business Media, 2012.

Tschichold, Jan. *The New Typography: A Handbook for Modern Designers*. Translated by Ruari McLean. Berkeley and Los Angeles: University of California Press, 1995.

Weibel, Peter, ed. *Beyond Art: A Third Culture: A Comparative Study in Cultures, Art and Science in 20th Century Austria and Hungary*. Berlin: Springer Science & Business Media, 2005.

Willen, Bruce and Nolen Strals. *Lettering & Type: Creating Letters and Designing Typefaces*. Princeton, NJ: Princeton Architectural Press, 2009.

9 The *sekai-sei* of circulation

The "world relevance" of the avant-garde Japanese calligraphy periodical *Bokubi* (1951–1960)[1]

Naomi Kuromiya

In Kyoto in June 1951, the calligrapher Morita Shiryū (1912–1988) published the inaugural issue of the monthly journal *Bokubi*. Its title, composed of the characters 墨 ["*boku*" or sumi ink] and 美 ["*bi*" or beauty], was telling: Morita envisioned *Bokubi* as an artistic research platform from which to explore new possibilities for Japanese calligraphy. Though publication of the periodical would continue until 1981, this study focuses on the extraordinary activity generated by circulation of *Bokubi* until roughly 1960. Through the intentionally international circulation of the magazine, Morita advocated for calligraphy's inclusion in the rapidly globalizing art world following World War II – a world connected in large part by the circulation of images and printed material. Several months later, in January 1952, Morita and several of his fellow experimental calligraphers officially founded the avant-garde calligraphy collective Bokujin-kai.[2] Dedicated to pushing the traditional art form into the modern era, the group began to publish reproductions of their works in *Bokubi*, and soon began organizing exhibitions of their work – first in Japan, and within a few years, abroad. By doing so, they intended to demonstrate the "*sekai-sei*," or "world relevance," of calligraphy to the international postwar art world.[3]

The locus of Bokujin-kai's activities, however, remained the group's journals, with *Bokubi* playing the most instrumental role in introducing the group to the world. The mission of the journal – promoting "modern" Japanese calligraphy – had a decidedly international outlook in its first decade, in turn presenting dozens of abstract Western artists working in a "calligraphic" style in Japan. Furthermore, despite being published almost exclusively in Japanese with a few captions or statements in French or English, *Bokubi* became one of the earliest postwar Japanese art periodicals to achieve an international circulation, broadcasting "modern" calligraphy and spurring further collaborative projects that introduced Bokujin-kai's works to Western art circles. *Bokubi* in these years functioned as a crucial transmitter of new images of (or related to) calligraphy, inspiring artists not only in Japan, but also in major modern art centers like New York and Paris.

Though there has been some examination of Bokujin-kai within the history of both postwar Western art and Japanese modern art, most scholarship has limited the discussion to the artistic influence of avant-garde calligraphy itself.[4] For reasons to be further discussed, despite its fascinating engagement with the wider art world from its inception into the 1960s, Bokujin-kai eventually faded from international artistic discussions, and the legacy of its mission both in Japan and in the West remains debatable. Certain elements of calligraphy arguably were absorbed into modern abstract painting practice and aesthetics, a theme addressed in several recent exhibitions on modern art.[5] But in terms of calligraphy itself as an avant-garde art form, there was a conspicuous international disinterest in its practice towards the end of the 1950s.

This chapter, therefore, reconsiders the impact of Bokujin-kai in its first decade through the material circulation of its journal *Bokubi*. Not only was the publication the major platform for Bokujin-kai works, and an earnest champion of calligraphy-related topics, but more important, the journal itself acted as an independent agent with an impactful role in creating international artistic connections. Capitalizing on a broad fascination with the concept and aesthetic of calligraphy in the 1950s, *Bokubi* presented it as the common denominator that helped facilitate some of the earliest postwar international conversations grounded specifically in modern art. Though not all these dialogues were fruitful, by examining the effects of the journal's reception in Japan and in Western countries, it becomes clear that those conversations' very existence – and Bokujin-kai's "world relevance" today – are largely indebted to *Bokubi*'s early, ambitious activities. The importance of the avant-garde calligraphy movement began, and continues to be founded in, the dialogues opened by the journal.

The launch of *Bokubi*

The timing of the first issue of *Bokubi* underscores its significance to the early history of Bokujin-kai: it was launched prior to the actual founding of the group, suggesting the understood value of a circulating magazine as a tool for publicizing avant-garde art – in this case, calligraphy. Many of the founding members had already been exploring new directions in calligraphy since the end of World War II, and through the American Occupation of Japan (1945–1952). Most were professionally trained in calligraphy.[6] One member – Hidai Nankoku (1912–1999) – was the son of the open-minded calligraphy master Hidai Tenrai (1872–1939), who already in the 1930s had aimed to rethink traditional values in calligraphy and integrate Western artistic concepts.[7] What Western observers might call "expressive" calligraphy, however, had a much longer heritage in Japan, dating back at least to the work of the Zen monk Hakuin Ekaku (1686–1768). When Morita began advocating for his and his colleagues' expressive works in 1947 with the publication *Sho no bi*,[8] there were certainly exciting formal innovations in their work, but what was more groundbreaking was Morita's advocacy for calligraphy as "avant-garde" with the potential for international recognition in the modern art world. *Sho no bi* was followed by *Bokubi*, and then *Bokujin* (launched in 1952).[9] The fact that Morita's earliest efforts focused on print circulation position the journal platform, as exemplified by *Bokubi*, as a key intellectual catalyst for the founding of the collective itself.

Calligraphy's position as an art form in Japan had been contested since the Westernization initiatives of the Meiji period (1868–1912). As Japan adopted Western models for art and art history, calligraphy was excluded in part because it did not fit into established Western categories for "high art."[10] Calligraphy was not included in the prestigious Japan Arts Exhibition [Nihon Bijutsu Tenrankai] until 1948.[11] Many future Bokujin-kai members, in contrast, felt that "art" and "calligraphy" need not be mutually exclusive. Their experiments ranged from the material – with the use of other media besides the traditional sumi ink and paper – to the formal – with the practice of increasingly gestural brushwork that rendered the written character [*kanji*] nearly illegible. The issue of legibility can also be traced to Meiji period debates: in 1882, the painter Koyama Shōtarō (1857–1916), who created Western-style oil paintings, published the incendiary article "Calligraphy Is Not an Art," claiming, "calligraphy is only a technology through which to write down the signs of a language . . . because it is to write characters with an established form, it leaves no room for creativity."[12] It is plausible that Morita and his colleagues saw a chance to modernize their calligraphy by making it a strikingly visual – rather than lexical – practice, and thus escape the debates of generations past.

The calligraphers' boundary-pushing experiments occurred at a time when the traditional arts were once again hotly debated. Calligraphy, due in part to its ideological links with the imperial wartime government, was outlawed from school education by the Occupation Authorities between 1947 and 1951.[13] At the same time, Japanese artists had begun revisiting prewar discussions of creating uniquely Japanese avant-garde art that might rival the creations of their Western counterparts – and, in doing so, allow them to participate in the growing "international" art community. The cultural environment, shaped by the rapid Westernization of the Occupation period, also triggered many artists to paradoxically seek inspiration for these new, unique "avant-garde" creations in Japanese traditional art, in part as a means of safeguarding Japanese art from complete Westernization. Just as "calligraphy" and "art" did not need to be separate fields, many artists felt that artistic practices could (and should) be both "Japanese" and "modern."[14] Reimagining calligraphy in the postwar era was not merely a fleeting visual interest but a matter of preserving Bokujin-kai members' artistic livelihood in the shifting cultural environment. To do so, they felt they needed to reposition calligraphy as "avant-garde" and artistically compelling.

The decision to form a group was, in and of itself, one strategy to align its members with other modern art collectives in both Japan and in the West. Not only had art collectives, or *dantai*, been a defining element of Japanese modern art, but Bokujin-kai could look to Western groups like Surrealism and Dada as models for advancing new theories in artistic practice.[15] Many avant-garde art groups declared themselves through a manifesto, and many advanced their manifesto through printed publications. In 1952, Bokujin-kai's manifesto, "Bokujin keisei aisatsu" ["Bokujin's Formative Greetings"], declared:

> European and American avant-garde artists and progressive critics and artists in Japan are knocking on the door of calligraphy. [. . .] Impoverished by

> its long hibernation in the shell of feudalism, this motion from without has started to shake calligraphy. We fully realize that we now stand at a pivotal moment: can the art of calligraphy, which has been guarding its long tradition in one corner of the Orient, revitalize itself as a true contemporary art, or will the very category [of calligraphy] become extinguished by being completely absorbed by progressive artists?[16]

They believed they could capitalize on a renewed interest in calligraphy, harnessing that energy to begin reforming the traditional art to fit the modern era, and eventually pushing calligraphy into a more influential role.

Morita's early mission for *Bokubi* aimed to accomplish this by establishing calligraphy as a unique art form with the capacity for international appreciation: he aimed for it to reach audiences outside of Japan who might grasp the "*sekai-sei*" of calligraphy.[17] *Bokubi*, and other projects seeking to reconsider traditional culture, participated in this dialogue on how to create and promote a uniquely "Japanese," yet internationally accepted, avant-garde art. A larger group known as the Gendai Bijutsu Kondankai (Contemporary Art Discussion Group), or "Genbi," united both traditional and modern artists: calligraphers, *ikebana* (flower arrangement artists) masters, potters, sculptors, and painters would gather to discuss how to "achieve world relevance through the innovation of Japanese tradition."[18] This is where Morita met some collaborators and contributors to *Bokubi* who hailed from outside the calligraphy discipline, including Yoshihara Jirō (1905–1972), the trained painter and future founder of the Gutai group that would pioneer abstract painting, installation art, and performance art in the later 1950s.[19]

The content of the early issues of *Bokubi* is a material testament to the group's complex and often conflicted interests: they at once wanted to publicize their artistic innovations within a traditional art, to preserve the traditional art form by characterizing it as avant-garde and internationally appreciated, and to create international modern art that was nonetheless uniquely Japanese. To this extent, we can see *Bokubi* both as a record of newly expressive calligraphy, and as the artists' earnest advocacy for its importance and comparability to other modern art practices. Western artists like the American painter Franz Kline (1910–1962), who created black forms on white grounds, particularly inspired Morita by fusing aesthetics of Japanese calligraphy with Western practice as a means of personal expression.[20] They also affirmed Morita's belief that calligraphy could be appreciated globally. In 1951 he posited: "if one can create forms that spring from the naked human heart, everyone will understand them without regard to time or place, race or occupation."[21] Fittingly, *Bokubi* was presented as a vehicle "for practical and theoretical research on calligraphic tradition, and for the exchange and clarification of cultures of the diverse countries of the world,"[22] while also ambiguously hinting at the challenges of this: "Rebellion and opposition – the path is thorny."[23] Throughout the 1950s, diverse content in the journal demonstrates that Morita hoped to strike a balance between these strong tensions.

The earliest issues of *Bokubi* embody this balance by reproducing (often within the same issue) Western abstract artworks, experimental Japanese calligraphy, and even traditional Japanese ink painting. Morita thus consciously honored the history of his art while presenting new directions for its evolution. The debut issue of *Bokubi* was clearly modern and international: it featured a painting by Kline on the cover and an essay on the state of modern art in postwar Japan vis-à-vis the West, with much of the text devoted to Kline and his paintings, by Japanese painter and writer Hasegawa Saburō (1906–1957).[24] Subsequent issues, however, would also include historic studies, such as a feature on ancient Japanese Buddhist statuary and its carved inscriptions in *Bokubi* no. 59 (October 1956). *Bokubi* no. 56 (July 1957) presented the miscellaneous calligraphy of Emperor Shōmu (reigned 724–749 AD), alongside articles on the sculpture and drawings of Alberto Giacometti (Figure 9.1).

In attempting to balance innovation and tradition in the publication's mission, *Bokubi* in its first ten years blurred these lines considerably, reflecting the dual interests inherent in an "avant-garde" traditional art form.[25] Avant-garde and tradition merged in Morita's monthly *Bokubi* feature dubbed the "*Arufa-bu*" ("Alpha Section"). Running from 1951 to 1953, it introduced completely illegible calligraphy – that is, works with no lexical referent whatsoever – by Bokujin-kai members including Inoue Yūichi (1986–1985) and Hidai Nankoku. Works by Western artists that resembled calligraphic works were notably *not* included in the "*Arufa-bu*." While there is no clear record of the reasons behind this, the restriction to only Japanese calligraphers may point to Morita's early or unconscious delineation of a separation between calligraphy and abstract painting.

Hasegawa, author of the feature essay in the debut issue and a frequent contributor, selected from calligraphers' submissions to the "*Arufa-bu*," and then provided his own commentary on each work. Perhaps Hasegawa assumed that, for a Japanese audience used to being able to read calligraphy, viewing these completely illegible works might be enhanced with some brief textual guidance. Hasegawa's selections and accompanying commentary ranged from "an extremely severe composition" (regarding a work by Morita composed of broad, broken brushstrokes in sharply angled, overlapping L-shapes) to "flexible, with freedom and strength" (regarding a work by Inoue featuring a cluster of bold, smooth, unbroken, curvilinear brushstrokes).[26] In general, Hasegawa's passionate voice and expertise in Western art was likely an important asset for legitimizing the magazine in Japan and abroad.[27]

The short-lived "*Arufa-bu*" represents some of the most radical, self-consciously "avant-garde" reproductions of works by Japanese calligraphers in *Bokubi*. These illustrations, as well as previously mentioned reproductions of Japanese and Western artworks, broke up the lengthy texts: among other articles, the journal printed transcripts of roundtable discussions between prominent art world figures, including Yoshihara (the future founder of Gutai) on the evolving role of calligraphy, directly recording the tension between safeguarding Japan and taking part in the international art dialogue. Texts such as these were printed only in Japanese, ostensibly for a Japanese audience, but often referenced material written by Western artists or recalled direct correspondence with Western colleagues. The journal's multivalent illustrations, meanwhile, were often captioned in English or French in

Figure 9.1 Cover of *Bokubi* no. 12, featuring *Painting No. 3* (1952) by Franz Kline, May 1952. Published by Shodō-shuppansha, Kyoto, Japan.

Source: Image courtesy The Metropolitan Museum of Art, New York, Thomas J. Watson Library.

addition to Japanese, thus opening the visual content to a more international group of consumers and collaborators. Occasionally, Morita's editor's note was also translated into English or French. The magazine also published Japanese translations of English or French texts by Western art figures like Willem de Kooning and William Stanley Hayter.[28]

Already in the early 1950s, then, *Bokubi* the magazine was beginning to realize some degree of "*sekai-sei*": it had become one of the earliest printed platforms to unite new works from the postwar art worlds of Japan and the West. While *Bokubi* was far from being Japan's first artistic connection with the West, it is a notable example of postwar international exchange that had a distinctly formal, often object-based, artistic focus. Kline's debut in *Bokubi* in 1951 was significant in that he was one of the first new Western artists to be introduced to Japan after the war, but beyond that, *Bokubi*'s presentation of Kline's affinities with calligraphy arguably made his work easier to evaluate and incited broader discussion within the Japanese art community, thereby launching the journal's impact *beyond* calligraphy.[29] Therefore, rather than simply "importing" and "exporting" artwork, *Bokubi* in its first years established modern artistic connections – and conversations – founded in formal aspects of calligraphy and calligraphic concepts as accessible starting points.

The circulation of *Bokubi* in Japan

The richest theoretical art discourse instigated by *Bokubi* occurred in Japan, largely due to the accessibility of both the images and the rich texts. By tying new, illegible calligraphy and new, abstract Western art to the recognizable tradition of calligraphy, *Bokubi* provided a strong comparative basis from which Japanese readers could analyze the "world relevance" of Bokujin-kai works. Furthermore, the extensive texts published in the 1950s dissected nuanced artistic issues that were not fully read or grasped by most American and French audiences. The texts and images together were thought-provoking, and as a result *Bokubi*'s readership seems to have steadily increased within the first year: after issue no. 12, in its second year of publication, the size of the periodical shrank from being printed on size B5 to A4 paper. This smaller size was likely more convenient to mail and suggests an expanded circulation after 1952. Despite the appetite for the contents of *Bokubi*, however, the fascination remained complicated, and it is possible to trace a concerted move in 1950s Japanese art circles to highlight modern art that moved *beyond* calligraphy – the conflicting goals of Bokujin-kai ultimately would become untenable within a public that continued to see calligraphy as too traditional.

There is certainly a clearly defined visual and performative legacy of calligraphy in Japanese modern art – in part because calligraphy has remained a consistently practiced art in Japan – but it is difficult to say whether certain expressive, "calligraphic" impulses in Japanese painting and performance in the 1960s (and beyond) were a result of the works and publicity of Bokujin-kai, or whether those impulses indicate outside inspiration from Western abstract painting.[30] It is likely a mix of both. Many paintings and performances by Gutai artists from the same period have either a spontaneous calligraphic quality or a lexical calligraphic quality, but those qualities could reflect either the known influence of Jackson Pollock's drip paintings on Yoshihara or inspiration from the Bokujin-kai group.[31] Regardless, the circulation of *Bokubi* around Japan pioneered conceptual questions that led to nuanced expansions in artistic practice and criticism. By theorizing possible new directions in postwar Japanese art (addressed through the lens of calligraphy), *Bokubi* pushed Japanese artists to think towards ambitious and experimental art that could maintain an international dialogue, as was later actualized by the Gutai group.

Unlike in Western countries, as will be outlined further below, for the Japanese art world, calligraphy had a centuries' old history that made it impossible to reduce it to a simple visual trope or newly spontaneous practice when discussing postwar art theory and possibilities.[32] As seen previously, the concept of calligraphy provided a solid foundation from which to absorb new Western art, like Kline's painting, and from which to facilitate communication on an international stage, but it did not guarantee that the new modern calligraphy works were themselves any easier to digest as avant-garde creations. Morita tried to balance international art with traditional practices and historic interests in *Bokubi*, but the reception of the magazine and the calligraphy it presented therein were equally torn between two negative extremes: either the works remained *too* tied to traditional calligraphy (and thus not groundbreaking enough to compete with European and American art), or, they dissolved the border between calligraphy and abstract art to the point of completely invalidating calligraphy as a practice – a practice that many remained invested in preserving. The first attitude was typical of abstract artists like Yoshihara; the second came to be the position, from around 1960 onward, of Morita and many Bokujin-kai calligraphers.

Bokubi hosted panels debating these issues as early as 1953, and the transcript of the roundtable "Calligraphy and Abstract Art" was published in *Bokubi* no. 26 (August 1953). Yoshihara, after much debate, made the following remarks, which foreshadow the formation of Gutai the following year, in 1954:

> [In abstract painting] it suffices for one mark to be a line and another a dot. I feel a great inexhaustible freedom in this. But when it comes to the syllable が [*ga*], there must be three dots and there are may prerequisites like this in calligraphy. [. . .] Shouldn't one transcend this limitation and devote oneself instead to form? [. . .] If the calligrapher raises standards to find that his calligraphy has become painting, what is wrong with that?[33]

が [*ga*] is a phonetic character in the Japanese alphabet – removing one dot or line would either change it to a different character, or make it illegible. Yoshihara, therefore, was arguing to push *beyond* calligraphy towards abstract painting – calligraphy seems to him to have been merely a stepping-stone (a parallel notion would arise abroad). Traces of Yoshihara's opinions at this panel discussion in *Bokubi* are visible in his 1956 Gutai Art Manifesto: "To today's consciousness,"

Yoshihara proclaimed, "the art of the past, which on the whole presents an alluring appearance, seems fraudulent."[34] This panel proves that calligraphy was spawning larger theoretical questions for Japanese artists seeking to create original works, and that *Bokubi* was circulating these debates around the Japanese art world. Yoshihara's participation in *Bokubi*'s discussions must have contributed to his subsequent decree to Gutai members to "create what has never been done before!"[35]

Even at the time of the "Calligraphy and Abstract Art" panel in 1953, when *Bokujin-kai* calligraphers were also publishing their most radical works in the "*Arufa-bu*," Morita remained hesitant about erasing the line between abstract painting and calligraphy. In the same discussion, Morita wondered whether the *moji-sei*, or "essence of the written character" was in fact the very foundation of calligraphy: "you may call this an over-simplification or a limitation of expression, but I think it is the characteristic of Eastern art, and it could mean that calligraphy is something that must be built on *moji-sei*."[36] A similar questioning of the value placed on the written character would surface in Morita's correspondence with Western colleagues, where he expressed concern with the "legibility" of the calligraphy to non-Japanese speakers, yet could not imagine discarding the character completely. This panel discussion in *Bokubi* clearly demonstrates an increasing tension between tradition and innovation in the Japanese art community of the 1950s and the problematic role of calligraphy in an art world that was attempting to become part of the international avant-garde.

The written character and its value became a point of contention even within the Bokujin-kai group, and one cannot help wondering if, had the members united in a common perspective on this matter, whether they might have been better equipped to fight for a place alongside abstract painting. Hidai Nankoku, for example, did not see the legibility or illegibility of his calligraphy as a relevant factor for appreciating his work. "I may not write characters, but it can be understood as calligraphy," he said.[37] The focus for him was on the beauty and function of the line: "I think what we call calligraphy exists because of the communication of the line."[38] Nankoku's colleague Inoue Yūichi experimented with nontraditional media, such as enamel painted on canvas with a handmade "broom brush" constructed from palm fronds. Like Nankoku, Yūichi's materials turned the focus from written characters to the texture of the line and its material interaction with the paper or canvas support. Yūichi, however, ultimately maintained the written character, and he spent the majority of his career producing calligraphy of single large characters [*ichijisho*] – although he did continue to experiment with "glue-ink," "frozen ink," and other unorthodox materials.[39]

Bert Winther-Tamaki has argued that Morita's growing insistence on maintaining a *moji-sei* fueled a conservative turn in Bokujin-kai and in the contents of *Bokubi* that slowly removed the group from the international art world.[40] Morita did ultimately find *moji-sei* to be the defining element of calligraphy. In an article published at the end of his career in 1992, Morita wrote: "Writing characters as a foundation, the throbbing of one's soul dances forth tied to a form: calligraphy is born where these two conditions are fulfilled."[41] In spite of any artistic parallels, international friendships, or "world relevance," Morita still believed calligraphy to have certain unique, essential characteristics. This mindset reveals that while Morita was conscious of the striking overlap between calligraphy and abstract painting, he saw those similarities – and the interest in them – as the basis from which to launch a reinvigoration of Japanese calligraphy, and in so doing, preserve the practice in modern times. Though he stated in 1951 that "before we are calligraphers, we must be artists; before we are artists, we must be human beings," Morita was less concerned with establishing a universal art for "human beings" and more so with convincing disparate art worlds that calligraphy was a vital component to human beings' existence.[42] This was the complex "world relevance" he had imagined for calligraphy.

In the short term, as the journal circulated abroad (and realized several major projects as a result), the recognition of *Bokubi* undoubtedly helped Bokujin-kai gain legitimacy as an international art group within Japan. Bokujin-kai exhibitions in Japan proliferated with the increased circulation of the journal. The first shows, beginning in 1953, were specifically dedicated to artists already published in the magazine, implying that *Bokubi* was the initial platform for introducing these artists to the public. Subsequent Bokujin-kai exhibitions, and other exhibitions that the members participated in in the 1950s, quickly grew to be equally as diverse as the contents of *Bokubi*: They included works by both Japanese and Western abstract artists, and became annual events frequented by the Japanese art world.[43] Just as *Bokubi* would encourage Western critics to consider modern art in new ways, its activity encouraged calligraphy exhibitions in Japan that provoked controversial questions from prominent Japanese art critics.

The legendary critic Takiguchi Shūzo reviewed the 1956 Bokujin-kai exhibition, damningly remarking on calligraphy's limitations as an expressive art, even when it managed to divorce itself from its inherent lexical referent:

> While this is an attempt to utilize the art of calligraphy as a vehicle for change, originality, and Eastern-style individuality, it would still, even after eliminating the ideograph, appear to be preserving a traditional standard for calligraphy. [. . .] To put it in more extreme terms, the potential for Japanese calligraphy cannot be located solely in its relationship with traditional paper, brush, and ink, and even expressions of space and form can and must be expressed in terms of our experience in the present. Ultimately, calligraphy that rejects the ideograph is itself a paradox, which, to put it in Western terms, can only result in another kind of sign.[44]

Takiguchi's remarks show how *Bokubi*'s activities managed to spur a fascination with calligraphy that launched questions about originality, individuality, and abstracted expression in Japanese art. Simultaneously, the works championed by *Bokubi*, meant to establish calligraphy's "*sekai-sei*" or "world relevance," proved to be too visually traditional for a Japanese audience. The conceptual potential for calligraphy lay in its treatment of space and form, rather than in the actual practice of calligraphy (what Takiguchi alludes to with "paper, brush, and ink"). For Takiguchi, a groundbreaking avant-garde thinker, even eliminating the *moji-sei* from Bokujin-kai's calligraphy would only create another rigid framework of abstracted "signs." Calligraphy itself thus morphed into the factor that limited Bokujin-kai's potential – fascinatingly, a parallel limitation would also arise in the Western art world. Still, *Bokubi* had clearly kick-started significant debates that would spur more radical projects of other art groups, most notably Gutai.

As the 1950s progressed, Morita would qualify his earlier, inclusive statements on the universality of form with more conservative declarations maintaining the absolute necessity of *moji-sei*. In *Bokubi* no. 87 (June 1959), Morita published a debate with Hidai Nankoku, where Morita emphasized the need to maintain calligraphy's traditional qualities: "To write the character, and in it to visually express yourself, that is what I think calligraphy is. I write characters to create calligraphy."[45] This traditional mindset had been foreshadowed in Takiguchi's review, and would only crystallize over time. Maruyama Kaori notes a turning point around 1960, shortly after the debate with Nankoku, where Morita began, without a clearly stated reason, to isolate his field from abstract painting when before he had embraced – even aggressively pushed – their overlap.[46]

By the 1960s, *Bokubi* would regularly dedicate entire issues to traditional or archaic calligraphy. *Bokubi* no. 177 (March 1968) was dedicated to Chinese illustrated manuscripts excavated from the Dunhuang caves, dating to the eighth or ninth century. When *Bokubi* ceased publication in 1981, its mission and audience were targeted primarily at the immediate calligraphy community. In a 2001 commemorative publication that reprinted a selection of significant articles from past issues of *Bokubi*, a section is devoted to contemporary Japanese calligraphers working very much in the style of Morita. No works by other contemporary Japanese artists were included, nor were works by comparable Western artists, suggesting a more insular audience than that of the 1950s.[47] This underscores the lasting significance of Morita's "inward" turn begun around 1960 – the very calligraphers that today follow in the Bokujin-kai calligraphers' path are not presented within an international artistic context.

And yet, when considering the group in a local, Japanese context, the heavy textual components of the early *Bokubi* issues reveal a complex, ongoing debate featuring provocative artistic questions that implicated a range of art world personalities. In the 1950s, Bokujin-kai's monthly efforts to record these debates and issues in *Bokubi* ensured that calligraphy was acknowledged as a relevant – if contentious – element of the postwar Japanese avant-garde dialogue. By recognizing the proactive nature of *Bokubi*, one cannot simplistically lump the journal together with calligraphy's overall ambiguous visual legacy in modern art. It was the questions Morita posed via *Bokubi* – rather than the actual calligraphic forms the magazine reproduced – that truly helped launch larger, "world relevant" art projects in Japan and beyond. As described above, *Bokubi*'s growing recognition abroad likewise seems to have helped Bokujin-kai garner prestige in Japan – in turn leading to even more debates and discussions referencing interactions with the West.

The circulation of *Bokubi* in New York and France

Bokubi's arrival in New York in the 1950s happened to coincide with a demonstrated interest in Japanese culture and art.[48] The Occupation had also already fostered travel and communication between the US and Japan. The Japanese American artist Isamu Noguchi was able to travel to Japan, and was instrumental in connecting the New York and Japanese art worlds. His mentor on his first postwar visit to Japan in 1950 was none other than Hasegawa. Noguchi was a key figure in the history of *Bokubi*: He brought with him reproductions of Kline's painting *Cardinal* (1950), which prompted Morita to feature Kline's work in the first issue of *Bokubi* (and inspired Hasegawa to write about Kline in his article in the inaugural issue).[49] This international artistic exchange, brought to an exciting fruition in *Bokubi*, inspired Kline to strike up a correspondence with Morita.

Kline sent a letter to Morita in August 1951, which Morita translated and reprinted in *Bokubi* no. 4, thus making their personal exchange evident to a wider audience. The letter also serves as an early positive review of the periodical, albeit biased, since Kline's own work was presented favorably. Kline enthusiastically wrote: "As soon as I received the translated text of Mr. Hasegawa's article about me in the first issue of *Bokubi*, I resolved to distribute *Bokubi* to every museum and publisher I know. I am confident that this perfect magazine will certainly be met with attention and interest."[50] Kline was true to his word. The following year, he wrote to Morita, who translated and published that letter in *Bokubi* no. 12. Kline stated: "I delivered several copies of *Bokubi* to Mr. Wittenborn and Mr. Egan. Both display them in their galleries . . ."[51] Kline's early enthusiasm for the publication was clear, and following his own recent successful shows in New York, his distribution of *Bokubi* to these prominent art world figures – including a publisher and a gallery owner – would have likely carried weight in further promoting the magazine in New York. Morita published his own reactions to Kline's work in the same issue (no. 12), in "On Looking at Mr. Kline's Latest Works: Impressions of a Calligrapher." In this way, via *Bokubi*, Morita and Kline started one of the earliest postwar artistic dialogues between the two countries.[52]

The reproductions of the works in *Bokubi* may have enhanced perceived aesthetic similarities between calligraphy and painting for both Morita and Kline – thus visually generating a productive, mutual exchange. It is a fortunate circumstance that both Kline's paintings and the calligraphy works were executed in black and white, and thus maintained more of their visual power in the black-and-white publication than color works would have. Unlike calligraphers who painted black ink onto white sheets of paper, however, Kline used white "positively."[53] "I paint the white as well as the black, and the white is just as important," Kline said.[54] The equal weight of black and white paints is nearly invisible in printed reproductions, where due to scale and print limitations a Kline canvas and a work of calligraphy could easily be mistaken for one another. Nonetheless, through these reproductions, *Bokubi*'s introduction of avant-garde calligraphy provided the basis for further artistic interactions. The uncanny resemblance between American painting and calligraphy became a fascinating point of comparison through which to begin and maintain artistic dialogue with Japan.

The French art community also expressed an interest in Japanese art after the war, at first regarding more traditional art forms, such as porcelain.[55] This interest was grounded in a longer history of fascination with Japanese visual culture, dating back at least to the nineteenth century appropriation of the woodblock print aesthetic.[56] As in New York, some of the earliest, postwar artistic communication between Japan and France can be traced to the concrete circulation of *Bokubi*, thus showcasing further evidence of *Bokubi*'s strength as a global communicator in the 1950s. A key promoter of *Bokubi* in France was the Belgian Cobra artist Pierre Alechinsky. Both Alechinsky and Morita recall starting a correspondence in 1952, in which Morita regularly sent him copies of *Bokubi*.[57] Like Kline in New York, Alechinsky seems to have enthusiastically recommended *Bokubi* to his French colleagues. Morita reprinted an article by Alechinsky in *Bokubi* no. 47, entitled "*Shūji ijō no mono*" ["Beyond writing"], in which Alechinsky compared calligraphy to the improvisational techniques of a jazz musician.[58] The recognizable aspects of calligraphy – for Alechinsky, its spontaneous, musical qualities – provided an accessible, artistically grounded basis for European artists to discuss art and calligraphy with Morita and the *Bokubi* audience.

These initial international connections were prominently featured in *Bokubi*, and subsequent high-profile projects instigated by these connections helped Bokujin-kai gain attention and recognition, especially at home in Japan. At the same time, many of these projects foreshadowed the limitations of calligraphy in an international context. These limitations would begin to reveal themselves over the course of the 1950s. This phenomenon is exemplified by an exhibition of works by Bokujin-kai members at The Museum of Modern Art in New York in 1954, entitled *Japanese Calligraphy* and organized with Morita's council. Morita's involvement seems to have been heavily linked to the production and circulation of *Bokubi*, copies of which he had sent to the exhibition's curator Arthur Drexler the year prior. It is also quite possible that the museum had received copies of *Bokubi* through New York galleries even earlier, prompting Drexler to connect with Morita – in any case, here again the magazine itself clearly spurred artistic exchange.

Morita assisted in the selection of works for the show by sending black-and-white photographs of Bokujin-kai calligraphy. He also attempted to provide Drexler with a theoretical basis for Bokujin-kai calligraphy and its significance in the context of modern art, writing in December 1953:

> Calligraphic works, too, seem to have come before the public as one of the artistic forms to be appreciated primarily for their visual beauty than as something to be read. Then I feel it is rather natural that we should lay more stress on the form which is to be seen and which is to express oneself than on the characters as things to be read, even though there is no cause to cast the latter entirely aside.[59]

This letter again recalls the Meiji argument that calligraphy was only writing, and Bokujinkai's shift to focusing on the visual power of calligraphy. Drexler, in his texts for the exhibition, picked up on the aesthetic connections between calligraphy and abstract painting: "The aesthetic development of calligraphy is aimed to appeal to the viewer even when the characters cannot be identified. It is interesting that such works show an approach towards the art of the abstract school of painting."[60]

Still, rather than articulating calligraphy as relevant international modern art in and of itself, Drexler posited that "the Japanese have successfully developed, out of their own traditions, an original contribution to modern painting."[61] Not only does this suggest that any Japanese painting prior to these works had not been "modern," but it also overlooks the entire premise of Bokujin-kai by classifying these new calligraphic works as "paintings." Drexler's exhibition texts in general hinted that only when calligraphy was freed from its traditional, lexical qualities, could it serve as inspiration for "original" modernist abstract painting as canonized by MoMA. This points to the value of calligraphy more as source material than as a modern art practice worth preserving.

A few years later, Morita also collaborated on a circulating exhibition, this time with Alechinsky in Europe. Entitled *L'encre de chine dans la calligraphie et l'art japonais contemporains* [*China ink in contemporary Japanese art and calligraphy*], it culminated its tour around Europe in Paris in 1956. The show was heavily featured in *Bokubi* no. 58 (September 1956). Unlike the MoMA exhibition, *L'encre de chine* presented more of the history of calligraphy, rather that just the formal qualities of modern calligraphy. Works by Bokujin-kai calligraphers appeared at the end of a chronological survey, situating them in a clear lineage. The exhibition's venue in Paris was the Musée Cernuschi, an Asian art museum, which further shifted the focus away from modern art, and also segregated the works within an East Asian history. The exhibition did use Bokujin-kai's innovations to pose broader artistic questions about the inherent freedom in abstract art versus writing.[62]

Despite successfully organizing several other exhibitions in both New York and Europe in subsequent years, *Bokubi's* critical reception abroad began to shift perceptibly over the course of the 1950s.[63] In New York, much of the exchange the journal stimulated was stifled due to a growing artistic nationalism, stemming in part from the critic Clement Greenberg's views. Greenberg's staunch rejection of any Japanese artistic connection was published in 1955, only shortly after the MoMA show and *Bokubi*'s initial circulation by Kline:

> none of the leading abstract expressionists except Kline has shown more than a cursory interest in Oriental art. . . . The fact that Far Eastern calligraphy is stripped and abstract – because it involves writing – does not suffice to make the resemblances to it in Abstract Expressionism more than a case of convergence. It is as though this country's possession of a Pacific coast offered a handy received idea with which to account for the otherwise inexplicable fact that it is now producing a body of art that some people regard as original.[64]

These comments reveal Greenberg's superficial understanding of calligraphy: he noted only its formal characteristics, without considering how other aspects of it – such as its temporal production – also fascinatingly "converged" with American abstract painting at that time. These statements also aligned with Greenberg's nationalist mission to promote the genius of American art.[65] Greenberg's rhetorical dodge of discussing calligraphy in any depth hints at the unequal relationship between American and Japanese art, and sensitivities over outside influence – to him, "convergences" may have been interesting, but not if they suggested a primary influence on American artists.

Greenberg's denial snowballed into a general denial of calligraphy.[66] Kline, despite being the first champion of *Bokubi*, began to distance himself from the journal beginning in 1956: "the artists [in Japan] decided that my work had its beginnings in Japanese calligraphy. . . . The whole subject is getting a little wearying," he said.[67] In 1962, he adamantly denied a connection with Japanese calligraphy: "calligraphy is writing, and I'm not writing."[68] These declarations make it difficult to ascertain whether any aspects of American Abstract Expressionism can be traced back to Bokujin-kai's calligraphy. The effects of Kline's statement surface in scholarship: Western inspirations for his "black-and-white abstractions" are analyzed, whereas Japanese calligraphy is briefly mentioned, but dismissed.[69] Scholarship on American art of this period tends to generalize the influence of calligraphy introduced by *Bokubi* by simply pointing to the lingering "calligraphic brushstroke" in American paintings.[70]

A declining interest in *Bokubi* in France occurred as a result the art critic Michel Tapié. The French artist Georges Mathieu introduced Tapié to the periodical, and in turn urged the critic to accompany him on a trip to Japan in 1957.[71] Morita published reproductions of Mathieu's works and translations of Tapié's articles in *Bokubi*, so they achieved a direct connection stimulated via the magazine.[72] However, similar to Yoshihara and Drexler, Mathieu and Tapié seem to have treated Bokujin-kai works more as inspirational springboards – in their case, for the discovery of "authentic" Japanese avant-garde artists who would fit into their ideal of international abstract painting.[73] Rather than denying any Japanese inspiration, as Greenberg had, Tapié and Mathieu instead openly appropriated elements of calligraphy as themes in their art or their criticism. Tapié saw calligraphy as just one aspect that composed his vision of an international art: he did not publish anything referencing Bokujin-kai, other than a brief statement that calligraphy had already been "assimilated as one international style of the avant-garde."[74] By 1957, Tapié was also receiving copies of the *Gutai* journal, which encouraged his decision to travel to Japan.[75]

This nonchalant "assimilation" is more obvious in Mathieu's activities. During his 1957 trip to Japan, Mathieu's work was exhibited at the Shirokiya department store in Tokyo. Before the show opened, Mathieu painted a mural (in what he imagined to be a "calligraphic" style) in the store window, wearing a kimono with his hair in a Japanese-style topknot. Of the performance, he claimed:

> Without knowing it, I gave these people, who have been seeking for nearly a century to combine the privileges of their traditional art with the seductions

> of Western painting, an answer to their question. I painted in oils, straight out of the tube, I used big flat brushes and long thin ones, I struck the canvas with folded towels dipped in liquid paint, I worked out the backgrounds by simply moving along the horizontal canvas.[76]

In the creation of one work, Mathieu was confident he had solved Bokujin-kai's dilemma, but he had clearly missed any of the subtle, complex issues being navigated by the group. Mathieu's comments specifically on *Bokubi*, meanwhile, reveal quite concretely that for Mathieu and for Tapié, the topic of calligraphy itself was too limited: "the preoccupations of Morita who edits *Bokubi* . . . show well how the most liberating intentions of the most audacious painter-calligraphers remain prisoners of ancient customs and habits."[77]

From a more positive standpoint, *Bokubi*'s initial role inspiring Tapié to visit Japan helped form his key relationship with Yoshihara and Gutai. When visiting Japan, Tapié found inspiration to reinvigorate his agenda to assemble an international group of *Informel* artists, and eventually promoted Yoshihara's Gutai as the Japanese contingent of *Informel*.[78] Bokujin-kai and its avant-garde aspirations for calligraphy were sidelined in the process. Winther-Tamaki contends that Morita's inward shift responded to "an enswampment of calligraphy by painting."[79] Perhaps because of this "enswampment," or simply because of the skyrocketing success of Gutai, Morita's inward turn also coincided with a conscious dismissal of calligraphy by art world figures like Takiguchi. In spite of its proactively international agenda, Bokujin-kai was disregarded because its works were ultimately still calligraphy, which was seen by the modern art world in all countries as a traditional art form that could not rid itself of its lexical constraints.

It is easy to treat Bokujin-kai's activities at best as a general precursor to more radical, more internationally recognized Japanese art, or, at worst, as a failure to realize their idealistic vision for calligraphy. These conclusions, however, overlook the significantly active role that *Bokubi* played both in Japan and internationally throughout Bokujin-kai's first decade. It is true that for artists and critics under Greenberg's purview, calligraphy became a limiting reagent to discussions of international influence, but Bokujin-kai's interactions with France (facilitated again through *Bokubi*), illuminate how the group's artistic influence lay in the material circulation of their journal and the dialogues and artistic connections it opened. Even in New York, the key relationships *Bokubi* established provided the foundation for later American exhibitions of Japanese art, such as MoMA's landmark 1966 exhibition *The New Japanese Painting and Sculpture*. The journal's "world relevance" is visible in an exhibition history that is a legacy of its international circulation.

Bokubi and its circulation as Bokujin-kai's legacy

Because of its shifting respect in various countries of the 1950s international art world, it is difficult to make conclusive statements about the legacy of Bokujin-kai in modern art. What is clear is that the group's journal, *Bokubi*, can be concretely linked to artistic connections that resulted in noteworthy collaborative projects. It established webs of influence connecting Japan, America, and Europe that are grounded in printed evidence. Like other avant-garde art periodicals before it, *Bokubi* was the key mode of transmitting the Bokujin-kai's new work and ideas around the world, and forged early postwar international artistic relationships. The mutually fascinating topic of calligraphy and its affinities with abstract painting constituted an artistically rooted topic of conversation that allowed for more fruitful international dialogue. These conversations resulted in collaborative exhibitions, which were rooted in images and ideas first published in the pages of *Bokubi*.

Bokubi necessarily raises a crucial question that may help reevaluate the role of printed material in scholarship: what does the perceived "failure" of Bokujin-kai's desired mission tell us about our own constricted narratives of modern art history? By examining *Bokubi* and its effects both in Japan and abroad in the 1950s, it is possible to see that Bokujin-kai's influence on modern art was ironically not rooted in the calligraphy it promoted, but in the larger questions and international projects catalyzed by what Kline had dubbed its "perfect" magazine. If one considers Bokujin-kai beyond its stated goals to make avant-garde calligraphy a global phenomenon, it is possible to understand that the exchanges and inspirations it fostered around the world through *Bokubi* endow the journal, rather than avant-garde calligraphy, with a traceable and lasting form of "world relevance." The brief but transcendent success of the circulation of a monthly periodical – one that fostered a network of exchange and led directly to several international exhibitions, eventually eclipsing the goals of an entire movement – is certainly not the "*sekai-sei*" that the Bokujin-kai calligraphers had envisioned as their legacy. But from today's standpoint, it nonetheless remains the clearest and most intriguing manifestation of their "world relevance."

Notes

1 Versions of this chapter were presented at the conferences *It Begins with a Story: Artists, Writers, and Periodicals in Asia*, Asia Art Archive and The University of Hong Kong, January 13, 2018; and at *Art, Institutions, and Internationalism: 1933–1966*, The Graduate Center, City University of New York, March 7, 2017. I am grateful for the feedback of participants and organizers at both conferences.

Unless otherwise noted, all translations are by the author.

2 The name *Bokujin-kai* is composed of three characters: 墨 ["*boku*" or sumi ink], 人 ["*jin*" or people], and 会 ["*kai*" or association]. The name has been translated as the "People of the Ink Group," "Association of People of Ink," and "The Human Ink Society."

3 Morita Shiryū, "*Henshū-shitsu yori*," *Bokubi* no. 2 (July 1951). Translated and reprinted in Alexandra Munroe, *Japanese Art after 1945: Scream against the Sky* (New York: H.N. Abrams, 1994), Exhibition catalogue, 373. The word "*sekai-sei*" could also be translated as "worldly essence" or "world character."

4 Such studies of Bokujin-kai include Munroe, *Japanese Art after 1945*; Bert Winther-Tamaki, "The Calligraphy and Pottery Worlds of Japan," in *Art in the Encounter of Nations: Japanese and American Artists in the Early Postwar Years* (Honolulu: University of Hawai'i Press, 2001), 66–109.

5 Notable recent shows have included *Black & White: Japanese Modern Art* at the Museum of Fine Arts, Boston (September 30, 2017–April 1, 2018) and *The Weight of Lightness: Ink Art* at M+ in Hong Kong (October 13, 2017–January 14, 2018).

6 For background on the group members, see Haryū Ichirō, "Sengo nihon no zeneisho: kaiga to no mitsugetsu jidai wo koete," in *Sho to kaiga to no atsuki jidai 1945–1969*, ed. O Art Museum (Shinagawa: Juridical Foundation and Cultural Promotion Association, 1992), Exhibition Catalogue, 3.

7 Stephen Addiss, "Japanese Calligraphy since 1968," in *Since Meiji: Perspectives on the Japanese Visual Arts, 1868–2000*, ed. Thomas J. Rimer (Honolulu: University of Hawai'i Press, 2012), 462.

8 *Sho no bi* [書の美] translates to "The Beauty of Calligraphy."

9 *Bokujin* [墨人], which is the same as "Bokujin" of "Bokujin-kai," translates to "Ink People" or "People of Ink."

10 See Addiss, "Japanese Calligraphy Since 1968," 445–446.

11 See Cecil B. Uyehara, *Japanese Calligraphy: A Bibliographic Study* (Lanham, MD: University Press of America, 1991), 11–13.

12 Ohno Kanako, "Japanese Calligraphy Is Not 'Art': The Meiji Government's Policy on Art and Japanese Calligraphy," *Ningenshakai kankyō kenkyū* no. 21 (2011), 40. Ohno also discusses the contested history of calligraphy as "art" in Meiji Japan, crucially discussing the roles of Ernest Fenellosa and Okakura Kakuzō (Tenshin).

13 For more on the contentious political position of calligraphy at this time, see Kenji Kajiya, "Modernized Differently: Avant-Garde Calligraphy and Art in Postwar Japan," Conference Paper, *M+ Matters: Postwar Abstraction in Japan, South Korea, and Taiwan*, Hong Kong, Hong Kong Arts Centre, June 28, 2014, 1–3.

14 For a more in-depth discussion of the contexts that gave rise to this movement, see Alexandra Munroe, "Avant-Garde Art in Postwar Japan: The Culture and Politics of Radical Critique 1951–1970" (PhD diss., Institute of Fine Arts, New York University, 2004), 143–152.

15 For more on art organizations in the history of Japanese modern art, see Alicia Volk, "Authority, Autonomy, and the Early Taishō Avant-Garde," *Positions: East Asia Cultures Critique* 21, no. 2 (Spring 2013): 451–473.

16 "*Bokujin-kai keisei aisatsu*" (1952) in *Bokujin 40-nen*, ed. Tsuji Futoshi et al. (Gifu City: Bokujin-kai, 1991), 84–85. Translated and reprinted in Bert Winther-Tamaki, *Art in the Encounter of Nations: Japanese and American Artists in the Early Postwar Years* (Honolulu: University of Hawai'i Press, 2001), 78.

17 Morita Shiryū, "*Henshū-shitsu yori*."

18 Munroe, *Japanese Art after 1945*, 87. Bokujin-kai members would also show calligraphy works in exhibitions organized by Genbi.

19 Yoshihara founded the Gutai Bijutsu Kyōkai (Gutai art association) in 1954, in Ashiya near Osaka. The journal *Gutai* was launched in 1955. Active until 1972, the group was infamous for radical painting and performances that broke all manner of artistic traditions. Members "leaped through paper screens, struggled against mud, wore dresses made of lightbulbs, flew paintings in the sky" [Ming Tiampo, "Please Draw Freely," in *Gutai: Splendid Playground*, ed. Alexandra Munroe and Ming Tiampo (New York: Guggenheim Museum, 2013), Exhibition catalogue, 45].

20 Franz Kline was active in New York in the 1940s and 1950s, and is today best known for his dramatic black-and-white abstract paintings in oil on canvas. He is associated with the American Abstract Expressionism movement, and considered one of its key figures.

21 Morita Shiryū, "*Henshū-shitsu yori*."

22 *L'encre de chine dans la calligraphie et l'art japonais contemporains* (Kyoto: Bokubisha, 1955), Exhibition catalogue, unpaginated.

23 See a translation of the "Mission of Bokubi" in Ellen Pearlman, *Nothing & Everything: The Influence of Buddhism on the American Avant Garde, 1942–1962* (Berkeley: Evolver Editions, 2012), 182.

24 The title of this article was "*Furansu to amerika kara no tayori: furui tōyō to atarashii seiyō toni kansuru zuisō*" ["News from France and America: Thoughts on the Old East and the New West"]. Hasegawa commented on artistic trends in the West that he observed at the first Salon de Mai in Tokyo (1950–1951), but devoted most of the text and the images to Kline.

25 These seemingly oppositional interests implicated other arts forms involved in the *Genbi* group, including pottery. For the pottery world's navigation of these issues, see Winther-Tamaki, *Art in the Encounter of Nations*, 89–105.

26 Hasegawa Saburō, "*Arufa bu*," *Bokubi* no. 13 (June 1952).

27 Winther-Tamaki notes that Hasegawa played a "leading role" in the Japanese art world, and had already begun before the war, in 1937, to bring "his extensive knowledge of pre-Meiji Japanese culture and European modern art to the task of forging channels of affinity" (Winther-Tamaki, *Art in the Encounter of Nations*, 33, 35).

28 A statement by De Kooning appeared in *Bokubi* no. 16 (September 1952). *Bokubi* no. 13 (June 1952) was a special issue devoted to Hayter. It included an article on Hayter by Hasegawa, and a translated artist's statement by Hayter.

29 Outside of this printed platform, there were several major exhibitions in Japan around the same time that introduced Western artists; for instance, the 1951 Third Yomiuri Independent show in Tokyo first exhibited works by Jackson Pollock. Pollock, however, notably had the most immediate impact on radical thinkers like Yoshihara. For more on this, see Lewis Kachur, "The View from the East: The Reception of Jackson Pollock among Japanese *Gutai* Artists," in *Abstract Expressionism: The International Context*, ed. Joan M. Marter (New Brunswick: Rutgers University Press, 2007), 152–153.

30 These questions are particularly evident in paintings by Shiraga Kazuo and Motonaga Sadamasa. See Gabriel Ritter, ed., *Between Action and the Unknown: The Art of Kazuo Shiraga and Sadamasa Motonaga* (New Haven: Yale University Press, 2015).

31 See Yamamoto Atsuo, "Some Viewpoints about Gutai," in *Art tribes: Galleria comunale d'arte moderna e contemporanea*, ed. Achille Bonita Oliva (Milan: Skira, 2002), 172.

32 The spontaneous, temporal quality of calligraphy was part of its traditional practice. All calligraphy is completed in one attempt: once a calligraphy piece is started, it is not possible for the calligrapher to go back and "fix" any errors. See Cecil H. Uyehara, "The Rite of Japanese Calligraphy and the Modern Age," *Oriental Art* 33 (Summer 1987): 175.

33 Originally published in *Bokubi* no. 26 (August 1953). Translated and cited by Winther-Tamaki, *Art in the Encounter of Nations*, p. 83.

34 Yoshihara Jirō, "Gutai Art Manifesto," in *Gutai: Splendid Playground*, trans. Reiko Tomii, 18–19.

35 Translated and cited in Ming Tiampo, "'Create What Has Never Been Done before!': Historicising Gutai Discourses of Originality," *Third Text* 21, no. 6 (November 2007): 689–706.

36 Arita Kōho, Morita Shiryū, Nakamura Makoto, Ōsawa Gakyū, Suda Kokuta, and Yoshihara Jirō, "Sho to Chūshō kaiga," *Bokubi* no. 26 (August 1953). Reprinted in *Bokubi kankō sanbyaku-ichigō, Bokujin gojūnen kinen, sho no utsukushisa* (Gifu-shi: Bokujinkai, 2001), 248.

37 Morita Shiryū and Hidai Nankoku, "Nani wo dō kangaete iru ka" ["What and How We Are Thinking"], *Bokubi* no. 87 (June 1959): 34.

38 Ibid., 37.

39 Takayuki Kurimoto, "The Art of Yūichi Inoue Examined through Three Terms," in *A Yuichi Inoue Retrospective: 1955–1985,* ed. Yuji Akimoto (Tokyo: Kamimori Foundation, 2016), 51.

40 See Winther-Tamaki, *Art in the Encounter of Nations*, 82–88.

41 Morita Shiryū, "Watashi ga mezashiteiru sho" ["The Calligraphy to Which I Aspire"], in *Sho to kaiga to no atsuki jidai 1945–1969*, ed. O Art Museum (Shinagawa: Juridical Foundation and Cultural Promotion Association, 1992), 136.
42 Shiryū, "*Henshū-shitsu yori.*"
43 The first of these exhibitions occurred in 1953, and was titled *Dai ikkai Bokujin kōbo shijō ten* [The first public exhibition of Bokujin based on the magazines] (*Bokubi kankō sanbyaku-ichigō*, 822).
44 Takiguchi Shūzō, "Higashi to nishi no sho," *Bokubi* no. 62 (January 1957). Translated by Maiko Behr and reprinted in *Primary Documents: From Postwar to Postmodern: Art in Japan 1945–1989*, ed. Doryun Chong (New York: The Museum of Modern Art, 2012), 77–78.
45 Shiryū and Nankoku, "Nani wo dō kangaete iru ka," 34.
46 Kaori Maruyama, "Sengo nihon ni okeru 'zenei sho' no tenkai: Morita Shiryū no sakuhin bunsetsu wo tōshite," in *Shogakushodōshi kenkyū* 23 (2013): 63–71.
47 See *Bokubi kankō sanbyaku-ichigō*, 772–821.
48 For example, there were noteworthy Japanese art exhibitions at the Marian Willard Gallery and at the Riverside Museum in 1954 (in collaboration with Hasegawa). The installation of a traditional Japanese house in the MoMA sculpture garden opened in the summers of 1954 and 1955, and lectures on Zen by D. T. Suzuki at Columbia University were attended by many artists. For an overview of this phenomenon, see Helen Westgeest, *Zen in the Fifties: Interaction in Art between East and West*, trans. Wendy van Os-Thompson (Zwolle: Waanders Publishers, 1996), 43–85.
49 Hasegawa recounts this in "*Furansu to Amerika kara no tayori.*" For more on Noguchi's relationship with Hasegawa, see Winther-Tamaki, *Art in the Encounter of Nations*, 33–39.
50 Franz Kline, "Amerika tsūshin" ["American Correspondence"], *Bokubi* no. 4 (September 1951). Reprinted in *Bokubi kankō sanbyaku-ichigō*, 366.
51 Franz Kline, "Henshū-sha e no tegami" ["Letter to the Editor"], *Bokubi* no. 12 (May 1952): 4. "Mr. Wittenborn" refers to the publisher George Wittenborn, who printed many artist periodicals in New York and was constantly in contact with artists via his bookstore. "Mr. Egan" refers to Charles Egan, Kline's dealer.
52 It was certainly not the first, as, for example, Noguchi and Hasegawa had connected a year earlier. In New York, it was also far from the only sign of interest in Japan (see note 47).
53 David Anfam, "Kline's Colliding Syntax: 'Black, White, and Things'," in *Franz Kline: Black and White*, ed. David Whitney (Houston: Houston Fine Art Press, 1995), Exhibition catalogue, 19.
54 Quoted in Katherine Kuh, *The Artist's Voice: Talks with Seventeen Artists* (New York: Harper and Row, 1962), 144.
55 As early as 1949, the head of the Musée Cernuschi in Paris, René Grousset, visited Japan and expressed interest in organizing an exhibition of Japanese porcelain in Paris (this was notably the museum which would later host a major Bokujin-kai exhibition). In the early 1950s, an exchange of exhibitions was organized between Tokyo and Paris, where an exhibition of French artists opened in Tokyo in February 1951; an exhibition of Japanese artists was included in the Salon de Mai in Paris in 1952. For more on this, see Segi Shin'ichi, *Sengo kūhakuki no bijutsu* (Tokyo: Shichō-sha, 1996), 60–68.
56 A comprehensive history of the French interest in Japan is traced in *Japonisme: The Japanese Influence on Western Art since 1858*, ed. Siegfried Wichmann (London: Thames & Hudson Ltd, 1981).
57 This is Morita's account as described in "Sho to Chūsho kaiga" ["Calligraphy and Abstract Art"], *Bokubi* no. 26 (August 1953). Reprinted in *Bokubi kankō sanbyaku-ichigō*, 243. See also Pierre Alechinsky, *Alechinsky: Paintings and Writings*, trans. Michael Fineburg (Pittsburgh: The Museum of Art, Carnegie Institute, 1977), 202.

58 Pierre Alechinsky, "Shūji ijō no mono" ["Beyond Writing"], *Bokubi* no. 47 (February 1955).
59 Morita Shiryū, letter to Arthur Drexler, December 12, 1953. The Museum of Modern Art Exhibition Records, 561.4. The Museum of Modern Art Archives, New York.
60 Arthur Drexler, "How to Appreciate Japanese Calligraphy." MoMA Exhs., 561.4. MoMA Archives, NY.
61 Arthur Drexler, "Abstract Japanese Calligraphy." MoMA Exhs., 561.4. MoMA Archives, NY.
62 Many of these questions implicated the postwar French artistic interest in exploring writing and signs. One example is Henri Michaux, although it also applied to many artists associated with Tapié's *Informel* movement. For Michaux's exploration of this topic, see Leslie Catherine Jones, "'A Barbarian in Asia': Henri Michaux's Works in Ink, 1927–1955" (PhD diss., Institute of Fine Arts, New York University, 2003), 38–71.
63 There were a number of other smaller exhibitions in both New York and Europe besides the two shows discussed here. Bokujin-kai had exhibitions at the Colette Allendy Gallery in Paris (1955), the Apollo Gallery in Brussels (1955), and the World House Gallery in New York (1957).
64 Clement Greenberg, "American Type Painting," in *Be-Bomb: The Transatlantic War of Images and All That Jazz, 1946–1956*, ed. Serge Guilbaut (Barcelona: MACBA and Museo Nacional Centro de Arte Reina Sofía, 2007), 652–653.
65 See Serge Guilbaut, "The New Adventures of the Avant-Garde in America: Greenberg, Pollock, or from Trotskyism to the New Liberalism of the 'Vital Center'," *October* 15 (Winter 1980): 61–78.
66 For an analysis of this phenomenon, see Bert Winther-Tamaki, "The Asian Dimensions of Postwar Abstract Art: Calligraphy and Metaphysics," in *The Third Mind: American Artists Contemplate Asia, 1860–1989*, ed. Alexandra Munroe (New York: The Solomon R. Guggenheim Foundation, 2009), Exhibition catalogue, 151–153.
67 Cited in Winther-Tamaki, *Art in the Encounter of Nations*, 58.
68 Ibid., 60.
69 For example, see Anfam, "Kline's Colliding Syntax: 'Black, White, and Things'," 10, 19.
70 The term "calligraphic brushstroke" was coined by William Seitz [Winther-Tamaki, "Calligraphy and Metaphysics," 149].
71 Alechinsky credits himself with this progression. It is possible that Mathieu had encountered *Bokubi* at an earlier date, but it is almost definite that Mathieu introduced it to Tapié. See Alechinsky, *Paintings and Writings*, 202–203.
72 For example, a translation of an article by Tapié entitled "*Kūkan to Hyōgen*" ["Emptiness and Expression"] appeared in *Bokubi* no. 65 (April 1957).
73 In 1962, Tapié published the catalogue *Avant-Garde Art in Japan* that featured artists he dubbed "truly creative individuals." Most were Gutai painters, including Motonaga Sadamasa, Shimamoto Shōzō, Shiraga Kazuo, Tanaka Atsuko, and Yoshihara. See Michel Tapié, *Avant-Garde Art in Japan* (New York: Harry N. Abrams, 1962), unpaginated.
74 Cited in Takiguchi, "Higashi to nishi no sho," 76.
75 Kawasaki Koichi, "Gutai," in *Art tribes: Galleria comunale d'arte moderna e contemporanea*, 180.
76 Georges Mathieu, *Au-delà du tachisme* (Paris: R. Juillard, 1963), 128. Translated and reprinted in Wichmann, ed., *Japonisme*, 406.
77 Mathieu, *Au-delà du tachisme*, 129.
78 Tapié's "canon" of *Informel* artists, including Kline, Tobey, Mathieu, Shimamoto, and Yoshihara appeared in Michel Tapié, *Une Morphologie Autre* (Torino, Italy: Edizioni d'Arte Fratelli Pozzo, 1960).
79 Winther-Tamaki, *Art in the Encounter of Nations*, 89. In an earlier statement, Morita had fretted about calligraphy "allowing itself to be cannibalized by progressive artists [i.e. noncalligraphers]" [Pearlman, *Nothing & Everything*, 181].

Bibliography

Addiss, Stephen. "Japanese Calligraphy since 1968." In *Since Meiji: Perspectives on the Japanese Visual Arts, 1868–2000*, edited by Thomas J. Rimer, 445–470. Honolulu: University of Hawai'i Press, 2012.

Alechinsky, Pierre. *Alechinsky: Paintings and Writings*. Translated by Michael Fineburg. Pittsburgh: The Museum of Art, Carnegie Institute, 1977.

Author Anon. *L'encre de chine dans la calligraphie et l'art japonais contemporains*. Kyoto: Bokubi Shuppansha, 1955. Exhibition catalogue.

Bokujinkai, ed. *Bokubi kankō sanbyaku-ichigō, Bokujin gojūnen kinen, sho no utsukushisa*. Gifu-shi: Bokujinkai, 2001.

Greenberg, Clement. "American Type Painting." In *Be-Bomb: The Transatlantic War of Images and All That Jazz, 1946–1956*, edited by Serge Guilbaut, 644–660. Barcelona: MACBA and Museo Nacional Centro de Arte Reina Sofía, 2007. Exhibition catalogue.

Guilbaut, Serge. "The New Adventures of the Avant-Garde in America: Greenberg, Pollock, or from Trotskyism to the New Liberalism of the 'Vital Center'." *October* 15 (Winter, 1980): 61–78.

Hasegawa, Saburō. "*Arufa bu*." *Bokubi* no. 13 (June 1952).

Jones, Leslie Catherine. "'A Barbarian in Asia': Henri Michaux's Works in Ink, 1927–1955." PhD diss., Institute of Fine Arts, New York University, 2003.

Kachur, Lewis. "The View from the East: The Reception of Jackson Pollock among Japanese *Gutai* Artists." In *Abstract Expressionism: The International Context*, edited by Joan M. Marter, 152–162. New Brunswick: Rutgers University Press, 2007.

Kajiya, Kenji. "Modernized Differently: Avant-Garde Calligraphy and Art in Postwar Japan." Conference Paper, M+ Matters: Postwar Abstraction in Japan, South Korea, and Taiwan, Hong Kong, Hong Kong Arts Centre, June 28, 2014.

Kawasaki, Koichi. "Gutai." In *Art tribes: Galleria comunale d'arte moderna e contemporanea*, edited by Achille Bonita Oliva, 145–151. Milan: Skira, 2002.

Kline, Franz. "Henshū-sha e no tegami" ("Letter to the Editor"). *Bokubi* no. 12 (May 1952).

Kuh, Katherine. *The Artist's Voice: Talks with Seventeen Artists*. New York: Harper and Row, 1962.

Kurimoto, Takayuki. "The Art of Yūichi Inoue Examined through Three Terms." In *A Yūichi Inoue Retrospective: 1955–1985*, edited by Yuji Akimoto, 46–54. Tokyo: Kami mori Foundation, 2016.

Maruyama, Kaori. "Sengo nihon ni okeru 'zenei sho' no tenkai: Morita Shiryū no sakuhin bunsetsu wo tōshite." *Shogakushodōshi kenkyū* 23 (2013): 63–71.

Mathieu, Georges. *Au-delà du tachisme*. Paris: R. Juillard, 1963.

Munroe, Alexandra. "Avant-Garde Art in Postwar Japan: The Culture and Politics of Radical Critique 1951–1970." PhD diss., Institute of Fine Arts, New York University, 2004.

Munroe, Alexandra. *Japanese Art after 1945: Scream against the Sky*. New York: H.N. Abrams, 1994. Exhibition catalogue.

The Museum of Modern Art Exhibition Records, 561.4. The Museum of Modern Art Archives, New York.

O Art Museum, ed. *Sho to kaiga to no atsuki jidai 1945–1969*. Shinagawa: Juridical Foundation and Cultural Promotion Association, 1992. Exhibition catalogue.

Ohno, Kanako. "Japanese Calligraphy Is Not 'Art': The Meiji Government's Policy on Art and Japanese Calligraphy." *Ningenshakai kankyō kenkyū* no. 21 (2011): 29–44.

Pearlman, Ellen. *Nothing & Everything: The Influence of Buddhism on the American Avant Garde, 1942–1962*. Berkeley: Evolver Editions, 2012.

Ritter, Gabriel, ed. *Between Action and the Unknown: The Art of Kazuo Shiraga and Sadamasa Motonaga*. New Haven: Yale University Press, 2015. Exhibition catalogue.

Segi, Shin'ichi. *Sengo kūhakuki no bijutsu*. Tokyo: Shichō-sha, 1996.

Shiryū, Moritaand Hidai Nankoku. "Nani wo dō kangaete iru ka" ("What and How We Are Thinking"). *Bokubi* no. 87 (June 1959).

Takiguchi, Shūzo. "Higashi to nishi no sho." *Bokubi* no. 62 (January 1957). Translated by Maiko Behr. Reprinted in *From Postwar to Postmodern: Art in Japan 1945–1989: Primary Documents*, edited by Doryun Chong, 76–78. New York: The Museum of Modern Art, 2012.

Tapié, Michel. *Avant-Garde Art in Japan*. New York: Harry N. Abrams, 1962.

Tapié, Michel. *Une Morphologie Autre*. Torino, Italy: Edizioni d'Arte Fratelli Pozzo, 1960.

Tiampo, Ming. "'Create What Has Never Been Done before!': Historicising Gutai Discourses of Originality." *Third Text* 21, no. 6 (November 2007): 689–706.

Tiampo, Ming. "Please Draw Freely." In *Gutai: Splendid Playground*, edited Alexandra Munroe and Ming Tiampo, 45–79. New York: Guggenheim Museum, 2013. Exhibition catalogue.

Uyehara, Cecil B. *Japanese Calligraphy: A Bibliographic Study*. Lanham, MD: University Press of America, 1991.

Uyehara, Cecil H. "The Rite of Japanese Calligraphy and the Modern Age." *Oriental Art* 33, no. 2 (Summer 1987): 174–182.

Volk, Alicia. "Authority, Autonomy, and the Early Taishō Avant-Garde." *Positions: East Asia Cultures Critique* 21 no. 2 (Spring 2013): 451–473.

Westgeest, Helen. *Zen in the Fifties: Interaction in Art between East and West*. Translated by Wendy van Os Thompson. Zwolle: Waanders Publishers, 1996.

Whitney, David, ed. *Franz Kline: Black and White*. Houston: Houston Fine Art Press, 1995. Exhibition catalogue.

Wichmann, Siegfried, ed. *Japonisme: The Japanese Influence on Western Art since 1858*. London: Thames & Hudson Ltd, 1981.

Winther-Tamaki, Bert. *Art in the Encounter of Nations: Japanese and American Artists in the Early Postwar Years*. Honolulu: University of Hawai'i Press, 2001.

Winther-Tamaki, Bert. "The Asian Dimensions of Postwar Abstract Art: Calligraphy and Metaphysics." In *The Third Mind: American Artists Contemplate Asia, 1860–1989*, edited by Alexandra Munroe, 145–157. New York: The Solomon R. Guggenheim Foundation, 2009. Exhibition catalogue.

Yamamoto, Atsuo. "Some Viewpoints about Gutai." In *Art tribes: Galleria comunale d'arte moderna e contemporanea*, edited by Achille Bonita Oliva, 153–197. Milan: Skira, 2002.

Yoshihara, Jirō. "Gutai Art Manifesto." In *Gutai: Splendid Playground*, edited by Alexandra Munroe and Ming Tiampo, translated by Reiko Tomii, 18–19. New York: Guggenheim Museum, 2013. Exhibition catalogue.

Index

Note: *Italic* page references indicate figures.

For Product Safety Concerns and Information please contact our EU representative GPSR@taylorandfrancis.com Taylor & Francis Verlag GmbH, Kaufingerstraße 24, 80331 München, Germany

Batch number: 10725973

Printed by Printforce, the Netherlands